AGRICULTURE ISSUES AND POLICIES

AGRICULTURAL CONSERVATION: PROGRAMS AND ISSUES

AGRICULTURE ISSUES AND POLICIES

Additional books in this series can be found on Nova's website under the Series tab.

Additional E-books in this series can be found on Nova's website under the E-book tab.

AGRICULTURE ISSUES AND POLICIES

AGRICULTURAL CONSERVATION: PROGRAMS AND ISSUES

STEVEN N. MURPHY
EDITOR

Nova Science Publishers, Inc.
New York

For permission to use material from this book please contact us:
Telephone 631-231-7269; Fax 631-231-8175
Web Site: http://www.novapublishers.com

Library of Congress Cataloging-in-Publication Data

Additional color graphics may be available in the e-book version of this book.

Agricultural conservation : programs and issues / editor: Steven N. Murphy.
p. cm.
Includes index.
ISBN 978-1-61761-288-6 (softcover)
1. Agricultural conservation--United States. 2. Conservation projects (Natural resources)--United States. I. Murphy, Steven N.
S604.6.A375 2010
333.72--dc22

2010027282

Published by Nova Science Publishers, Inc. † New York

CONTENTS

PREFACE

The U.S. Department of Agriculture provides technical and financial assistance to attract interest and encourage participation in conservation programs. Participation in all USDA conservation programs is voluntary. These programs protect soil, water, wildlife, and other natural resources on privately owned agricultural lands to limit environmental impacts of production activities both on and off the farm, while maintaining or improving production of food and fiber. Some of these programs center on improving or restoring resources that have been degraded, while others create conditions to limit degradation in the future. This book explores the numerous programs and issues currently established by the government for agricultural conservation and how improved management controls of these programs can enhance the effectiveness of conservation

Chapter 1- The Natural Resources Conservation Service (NRCS) and the Farm Service Agency (FSA) in the U.S. Department of Agriculture (USDA) currently administer over 20 programs and subprograms that are directly or indirectly available to assist producers and landowners who wish to practice conservation on agricultural lands. The number, scope, and overall funding of these programs has grown in recent years. This growth can cause some confusion over which problems and conditions each program addresses, and specific program characteristics and performance. The programs are as follows:

- Agricultural Management Assistance (AMA) Program
- Chesapeake Bay Watershed Program
- Conservation Operations (CO); Conservation Technical Assistance (CTA)
- Conservation Reserve Program (CRP)

- CRP—Conservation Reserve Enhancement Program (CREP)
- CRP—Farmable Wetlands Program
- Conservation Security Program
- Conservation Stewardship Program (CSP)
- Emergency Conservation Program (ECP)
- Emergency Watershed Program (EWP)
- Environmental Quality Incentives Program (EQIP)
- EQIP—Agricultural Water Enhancement Program (AWEP)
- EQIP—Conservation Innovation Grants (CIG)
- Farmland Protection Program (FPP)
- Grassland Reserve Program (GRP)
- Healthy Forest Reserve Program (HFRP)
- Resource Conservation and Development (RC&D) Program
- Watershed and Flood Prevention Operations
- Watershed Rehabilitation Program
- Wetland Reserve Program (WRP)
- Wildlife Habitat Incentive Program (WHIP)

This tabular presentation provides basic information introducing each of the programs. In each case, a brief program description is followed by information on major amendments in the 2008 farm bill (P.L. 110-246); national scope and availability; states with the greatest participation; the backlog of applications or other measures of continuing interest; program funding authority; FY2010 estimated spending; the FY2011 Administration budget request; statutory authority; the authorization expiration date; and a link to the program's website.

Chapter 2- Agricultural conservation has been a public policy issue for more than 60 years. Congress has repeatedly taken action on the issue through water and soil legislation, often as part of omnibus farm bills. Early policy decisions were directed at addressing natural resource concerns on the farm, primarily reducing high levels of soil erosion and providing water to agriculture in quantities and quality that enhanced farm production. In more recent years, conservation policy has shifted to concerns about the off-farm impacts of agricultural activities.

The latest farm bill, the Food, Conservation, and Energy Act of 2008 (P.L. 110-246), reauthorized most existing conservation programs, modified several programs, and created a few new programs. The U.S. Department of Agriculture's (USDA's) conservation efforts have centered on implementing these conservation programs through working land conservation practices, retiring land from production or establishing conservation easements, and providing technical assistance. Program implementation controversies could lead to congressional oversight or action, especially given the recent financial statement audit reports on conservation program payments at USDA.

Chapter 3- The U.S. Department of Agriculture (USDA) administers conservation programs, such as the Conservation Stewardship Program (CSP, formerly the Conservation Security Program) and the Environmental Quality Incentives Program (EQIP), to help farmers reduce soil erosion, enhance water supply and quality, and increase wildlife habitat, among other things.

This testimony is based on GAO reports on CSP and EQIP, each issued in 2006, and a 2008 report on farm program payments. It discusses (1) the potential for duplicate payments between CSP and other conservation programs, (2) USDA's process for allocating EQIP funds to the states to optimize environmental benefits, and (3) USDA's management controls over farm program payments.

Chapter 4- Mr. Chairman and members of the subcommittee, we appreciate the opportunity to review conservation programs delivered by the U.S. Department of Agriculture (USDA). We are pleased to share our experiences in implementing the Conservation Title. We will also offer our observations on the changing business environment in which programs operate, the working relationships with our USDA conservation partners, and the opportunities and challenges we face in implementing the 2008 Farm Bill.

In: Agricultural Conservation: Programs and Issues ISBN: 978-1-61761-288-6
Editors: Steven N. Murphy

Chapter 1

AGRICULTURAL CONSERVATION: A GUIDE TO PROGRAMS

Megan Stubbs

SUMMARY

The Natural Resources Conservation Service (NRCS) and the Farm Service Agency (FSA) in the U.S. Department of Agriculture (USDA) currently administer over 20 programs and subprograms that are directly or indirectly available to assist producers and landowners who wish to practice conservation on agricultural lands. The number, scope, and overall funding of these programs has grown in recent years. This growth can cause some confusion over which problems and conditions each program addresses, and specific program characteristics and performance. The programs are as follows:

- Agricultural Management Assistance (AMA) Program
- Chesapeake Bay Watershed Program
- Conservation Operations (CO); Conservation Technical Assistance (CTA)
- Conservation Reserve Program (CRP)
- CRP—Conservation Reserve Enhancement Program (CREP)
- CRP—Farmable Wetlands Program

- Conservation Security Program
- Conservation Stewardship Program (CSP)
- Emergency Conservation Program (ECP)
- Emergency Watershed Program (EWP)
- Environmental Quality Incentives Program (EQIP)
- EQIP—Agricultural Water Enhancement Program (AWEP)
- EQIP—Conservation Innovation Grants (CIG)
- Farmland Protection Program (FPP)
- Grassland Reserve Program (GRP)
- Healthy Forest Reserve Program (HFRP)
- Resource Conservation and Development (RC&D) Program
- Watershed and Flood Prevention Operations
- Watershed Rehabilitation Program
- Wetland Reserve Program (WRP)
- Wildlife Habitat Incentive Program (WHIP)

This tabular presentation provides basic information introducing each of the programs. In each case, a brief program description is followed by information on major amendments in the 2008 farm bill (P.L. 110-246); national scope and availability; states with the greatest participation; the backlog of applications or other measures of continuing interest; program funding authority; FY2010 estimated spending; the FY2011 Administration budget request; statutory authority; the authorization expiration date; and a link to the program's website.

INTRODUCTION

Currently, more than 20 U.S. Department of Agriculture (USDA) agricultural conservation programs exist to assist private landowners with natural resource concerns. Though some similarities among these programs exist, each is administered with slight differences. These differences and the number of programs create some general confusion about the purpose, participation, and policies of the programs.

The number of agricultural conservation programs has steadily increased over the past 60 years. Early conservation efforts were focused on reducing high levels of soil erosion and providing water to agriculture in quantities and quality that enhanced farm production. Congress responded to these issues by creating and revising programs designed to reduce resource problems on the farm. By the early 1980s, however, concern was growing that these programs were not adequately dealing with environmental problems resulting from agricultural activities (especially off the farm). In 1985, conservation policy took a new direction when Congress passed the Food Security Act of 1985 (1985 farm bill, P.L. 99-198), which established the first conservation programs designed to deal with environmental issues resulting from agricultural activities.

Provisions enacted in subsequent farm bills, including in 1990, 1996, 2002, and 2008,[1] reflect a rapid evolution of the conservation agenda, including the growing influence of environmentalists and other non-agricultural interests in the formulation of conservation policy, and a recognition that agriculture was not treated like other business sectors in many environmental laws. Congress also began funding many of these new programs through mandatory spending for the first time, using the borrowing authority of USDA's Commodity Credit Corporation (CCC)[2] as the funding mechanism instead of annual appropriations. In addition to the original soil erosion and water quality and quantity issues, the conservation agenda has continued to expand to address other natural resource concerns, such as wildlife habitat, air quality, wetlands restoration and protection, energy efficiency, and sustainable agriculture.

Lead agricultural conservation agencies within USDA are the Natural Resources Conservation Service (NRCS), which provides technical assistance and administers most conservation programs, and the Farm Service Agency (FSA), which administers the largest program, the Conservation Reserve Program (CRP). These agencies are supported by others in USDA that supply research and educational assistance, including the Agricultural Research Service (ARS), the Economic Research Service (ERS), and the Forest Service (FS).[3] In addition, the conservation effort involves a large array of partners, including other federal agencies, state and local governments, and private organizations, among others, who provide funds, expertise, and other forms of assistance to the conservation effort.

OVERVIEW

USDA provides technical and financial assistance to attract interest and encourage participation in conservation programs. Participation in all USDA conservation programs is voluntary. These programs protect soil, water, wildlife, and other natural resources on privately owned agricultural lands to limit environmental impacts of production activities both on and off the farm, while maintaining or improving production of food and fiber. Some of these programs center on improving or restoring resources that have been degraded, while others create conditions to limit degradation in the future.

Though programs in this report are listed alphabetically, they can be grouped into the following categories based on similarities: working land programs, land retirement and easement programs, watershed programs, emergency programs, compliance and technical assistance programs, and other. **Figure 1** illustrates the relative size of each group based on the amount of funding in FY2009.

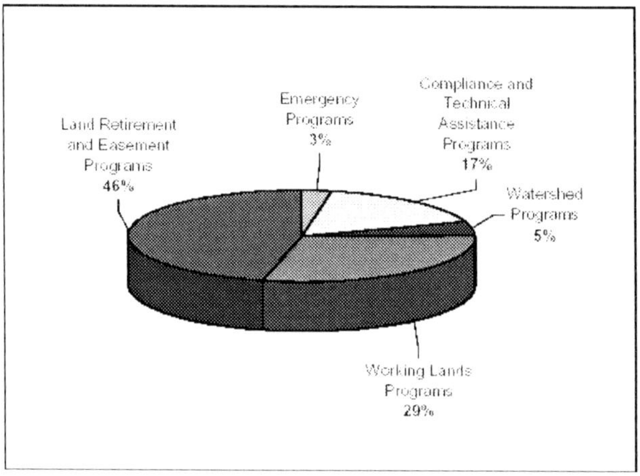

Source: CRS Funding estimates compiled from FY2009 Omnibus Appropriations Act (P.L. 111-8), American Recovery and Reinvestment Act of 2009 (P.L. 111-5), and USDA, FSA, *CCC Commodity Estimates Book*, FY2010 Mid-Session Review, August 25, 2009, http://www.fsa.usda.gov/FSA/webapp? area=about&subject = landing & topic= bap-bu-cc.

Notes: Technical assistance includes funding for compliance programs. Funding for extension and research is not included. Total funding for conservation programs is approximately $5.3 billion.

Figure 1. FY2009 Funding for Conservation Programs

The majority of conservation programs are funded through USDA's Commodity Credit Corporation (CCC) as mandatory spending. Congress authorizes mandatory programs at specified funding levels each year (or acreage enrollment levels for the Conservation Reserve, Conservation Stewardship, Wetlands Reserve, and Grasslands Reserve Programs). They are funded at these levels unless Congress limits funding to a lower amount through the appropriations or legislative process (or puts a ceiling on acreage that can be enrolled). Discretionary programs are funded each year through the annual appropriations process.

Despite a steady increase in mandatory funding authority, many conservation programs have been reduced or capped through annual appropriations acts since FY2003. Many of these spending reductions were at the request of the Bush Administration, though the Obama Administration has also requested reductions in the FY2011 proposal. The mix of programs and amounts of reduction has varied from year to year. Some programs, such as the CRP, have not been reduced by appropriators in recent years, while others, such as the Environmental Quality Incentives Program (EQIP), have been repeatedly reduced below authorized levels. EQIP's authorized mandatory funding was reduced by over $1.16 billion from FY2005 through FY2010. Even with these reductions, total mandatory funding for conservation programs has grown from a total of $3.6 billion in FY2005 to a total of $4.7 billion in FY2010.

Conservation Programs

The tabular presentation that follows provides basic information introducing each of the USDA agricultural conservation programs, including:

- administering agency or agencies within USDA;
- a brief program description;
- major amendments to the program in the Food, Conservation, and Energy Act of 2008 (P.L. 110-246), commonly referred to as the 2008 farm bill;
- national scope and availability, including participation levels and acres enrolled where available;
- states with the highest level of funds obligated or acres enrolled (where applicable);
- the volume of application backlog or public interest in each program;
- the authorized funding levels, whether it is mandatory spending or discretionary appropriations, and any funding restrictions where applicable;

- the FY2010 funding level provided by the Agriculture, Rural Development, Food and Drug Administration, and Related Agencies Appropriations Act of 2010 (P.L. 111-80) or, if applicable, the Food, Conservation, and Energy Act of 2008 (P.L. 110-246) authorized level;
- the FY2011 funding level requested by the Administration;
- statutory authority, recent amendments, and U.S. Code reference;
- the expiration date of program authority unless permanently authorized; and
- a link to the program's website.

Information for the following tables is drawn from agency budget presentations, explanatory notes, and websites; written responses to questions published each year in hearing records of the Agriculture Appropriations Subcommittees of the House and Senate Appropriations Committees; and spending estimates from the Congressional Budget Office. Further information about these programs may be found on the NRCS website at http://www.nrcs.usda.gov and on the "conservation programs" page of the FSA website at http://www.fsa.usda.gov. For additional information about agricultural conservation in general, see CRS Report R40692, *Agricultural Conservation Issues in the 111th Congress*.

USDA Agricultural Conservation Programs

Working Lands Programs—typically classified as programs that allow private land to remain in production, while implementing various conservation practices to address natural resource concerns specific to the area.

- Environmental Quality Incentives Program (EQIP), Conservation Stewardship Program (CSP)/Conservation Security Program, Wildlife Habitat Incentives Program (WHIP), Agricultural Water Enhancement Program (AWEP), and Agricultural Management Assistance (AMA).

Land Retirement and Easement Programs—land retirement programs provide federal payments to private agricultural landowners for *temporary* changes in land use or management to achieve environmental benefits. Conversely, conservation easements impose a *permanent* land-use restriction that is voluntarily placed on the land in exchange for a government payment.

- Conservation Reserve Program (includes the Conservation Reserve Enhancement Program (CREP) and Farmable Wetlands), Wetlands Reserve Program (WRP), Farmland Protection Program (FPP), Grassland Reserve Program (GRP), and Healthy Forests Reserve Program (HFRP).

Watershed Programs—NRCS partners with local sponsors to carry out activities for soil conservation; flood prevention; conservation, development, utilization and disposal of water; watershed surveys; and dam and flood structure rehabilitation.

- Watershed and Flood Prevention Operations (also referred to as the Small Watershed Programs, P.L. 566 and P.L. 534), Watershed Rehabilitation, and Watershed Surveys and Planning.

Emergency Programs—provide disaster assistance for farmland rehabilitation and impairments to watersheds. Programs are usually funded through supplemental appropriation acts.

- Emergency Conservation Program (ECP) and Emergency Watershed Program (EWP) (includes floodplain easements).

Compliance—prohibits a producer from receiving many federal farm program benefits (including conservation assistance) when conservation program requirements for highly erodible lands and wetlands are not met.

- Conservation Compliance, Sodbuster, Swampbuster, and Sodsaver.

Technical Assistance Programs—provides landowners with science-based conservation information and technical expertise (e.g., engineering and biological) unique to the region and land use type. Usually does not include financial assistance.

- Conservation Operations (includes Conservation Technical Assistance, Survey, Soil Survey, Grazing Lands Conservation Initiative, and Plant Materials Centers), Resource Conservation and Development (RC&D) program.

Other Conservation Programs and Provisions—Chesapeake Bay Watershed Program, Cooperative Conservation Partnership Initiative, Conservation Innovation Grants, Great Lakes Basin Program, Regional Equity, State Technical Committees, Voluntary Public Access and Habitat Incentive Program, and Grassroots Source Water Protection Program.

Agricultural Management Assistance (AMA) Program

Administering Agency(s)	NRCS (conservation assistance), Agricultural Marketing Service (AMS, organic certification), and Risk Management Agency (RMA, production, price, or revenue risk reduction)
Program Description	AMA provides cost-sharing assistance unde contracts of 1 to 10 years to producers in 16 specified states where participation in the federal crop insurance program has been historically low. Producers use this assistance to construct or improve water management and irrigation structures, plant trees, control soil erosion, practice integrated pest management, practice organic farming, develop value-added processing, and enter into futures, hedging, or options contracts to reduce production, price, or revenue risk.
Major 2008 farm bill amendments	Adds Hawaii as an eligible state; authorizes funding of $15 million annually from FY2008 through FY2012, and $10 million in subsequent years; and designates program funding for USDA agencies: 50% - NRCS, 10% - AMS, and 40% - RMA.
National Scope	Not available nationwide. Eligible states include: CT, DE, HI, MD, MA, ME, NV, NH, NJ, NY, PA, RI, UT, VT, WV, and WY. As of the end of FY2009, 723 contracts are being implemented within these states.

Leading States	States with the most funds obligated in FY2009 include PA ($1.3 million), NY ($1.1 million), and WY ($0.9 million).
Backlog/Interest	A backlog of 266 applications was pending at the end of FY2009, up from a backlog of 94 applications in FY2008. These applications would enroll more than 23,471 acres at a cost of $10.3 million.
Funding authority	Mandatory. Permanently authorized at $10 million for each fiscal year except FY2008-FY2012, which is authorized at $15 million for each fiscal year. Funding is split by law between the three USDA agencies: 50% - NRCS, 10% - AMS, and 40% - RMA.
FY2010 funding	$15 million ($7.5 million for conservation).
FY2011 Administration request	$10 million ($2.5 million for conservation).
Statutory authority	Authorized in Title I, §133 of the Agricultural Risk Protection Act of 2000 (P.L. 106-224) as §524(b) of the Federal Crop Insurance Act, as amended. Amended in §2801 of the Food, Conservation, and Energy Act of 2008 (P.L. 110-246). 7 U.S.C. 1524(b).
Authorization expires	Permanent authorization.
Program website	http://www.nrcs.usda.gov/ programs/ AMA/ index.html

Chesapeake Bay Watershed Program

Administering Agency(s)	NRCS
Program Description	The Chesapeake Bay Watershed Program provides additional financial assistance through existing conservation programs for areas within the Chesapeake Bay watershed. Law requires special consideration for producers in the Susquehanna, Shenandoah, Potomac, and Patuxent river basins. Program rules, contracts, and sign up depend on the program through which NRCS is providing the funds. In a notice published in the *Federal Register* on March 12, 2010 (75 FR 11832), NRCS announced that FY2010 funds are provided through the Environmental Quality Incentives Program (EQIP) and the Cooperative Conservation Partnership Initiative (CCPI), which includes EQIP and the Wildlife Habitat Incentive Program (WHIP). EQIP and WHIP are discussed further below.
Major 2008 farm bill amendments	Program created in the 2008 farm bill.
National Scope	Not available nationwide. The program applies to all tributaries, backwaters, and side channels, including watersheds draining into the Chesapeake Bay. This includes areas within DE, MD, NY, PA, VA, and WV.
Leading States	In FY2009, states with the most funding obligated were PA ($6.7 million), VA ($6.2 million), and MD ($5 million).
Backlog/Interest	None identified.

Funding authority	Mandatory. FY2009 - $23 million; FY2010 - $43 million; FY2011 - $72 million; and FY2012 -$50 million.
FY2010 funding	$43 million.
FY2011 Administration request	$72 million.
Statutory authority	Authorized in Title II, §2605 of the Food, Conservation, and Energy Act of 2008 (P.L. 110-246) as §1240Q of the Food Security Act of 1985 (P.L. 99-198). 16 U.S.C. 3839bb-4.
Authorization expires	September 30, 2012.
Program website	http://www.nrcs.usda.gov/feature/chesapeakebay/chesapeakebay.html

Conservation Operations (CO)—Conservation Technical Assistance (CTA)

Administering Agency(s)	NRCS
Program Description	Conservation Operations (CO) is the primary account funding technical assistance within NRCS. More than 80% of CO funding is for Conservation Technical Assistance (CTA), which provides conservation planning and implementation assistance through field staff placed in almost all counties within the United States and territories. This assistance is provided to producers and land owners who voluntarily apply natural resource conservation systems, consisting of one or more practices, on private and other non-federal lands. Other

	components of CO include the Grazing Lands Conservation Initiative, soil surveys, snow surveys, and plant material centers.
Major 2008 farm bill amendments	Amends the Soil Conservation and Domestic Allotment Act include a definition of technical assistance and agriculture commodity. Other provisions indirectly affect this program, including allowing NRCS to approve qualified individuals and entities, referred to as third parties, to provide some types of technical assistance and retaining a cap on total funding for technical assistance provided through mandatory programs.
National Scope	Available nationwide. CTA was funded at $790.9 million and 5,624 staff years for FY2009, according to the FY2011 budget notes. Total CO spending for FY2009 was $887.6 million and 6,573 staff years.
Leading States	No data are available for the CTA subset in FY2010, but the three leading stats for total CO funding (estimate) are: TX ($47.3 million), IA ($23.7 million), and KS ($23.1 million).
Backlog/Interest	Not available.
Funding authority	Discretionary. No amount specified.
FY2010 funding	$756 million for CTA out of $887.6 million for all CO.
FY2011 Administration request	$796.4 million for CTA out of $923.7 million for all CO.
Statutory authority	Authorized in the Soil Conservation and Domestic Allotment Act (P.L. 74-46), as amended.

	Amended in §2802 of the Food, Conservation, and Energy Act of 2008 (P.L. 110-246). 16 U.S.C. 590a-g, 16 U.S.C. 590q.
Authorization expires	Permanent authorization
Program website	http://www.nrcs.usda.gov/programs/cta/ (CTA); http://www.glti.nrcs.usda.gov/ (grazing); http://soils.usda.gov/ (soil survey); http://www.wcc.nrcs.usda.gov/ (snow survey); and http://plant-materials.nrcs.usda.gov/ (plant materials).

Conservation Reserve Program (CRP)

Administering Agency(s)	FSA, technical assistance by NRCS
Program Description	CRP provides annual rental payments, usually over 10 years, to producers to replace crops on highly erodible and environmentally sensitive land with long-term resource conserving plantings. Bids to enroll land are solicited during a limited time period, then compared using an Environmental Benefits Index (EBI). Those with the highest EBI scores are accepted. Imbedded in the CRP are several small and more focused programs that bypass the general bidding process, some established in law and others established administratively, to address specific resource topics, including more concentrated resource problems in a portion of a state, protection of small isolated agricultural wetlands, and improvement of habitat for upland game birds. All lands that qualify for these subprograms are automatically accepted.

Major 2008 farm bill amendments	Reduces enrollment ceiling from 39.2 million acres to 32 million acres; amends haying and grazing conditions; and authorizes incentives to assist socially disadvantaged and beginning farmers.
National Scope	Available nationwide. According to the FY2011 budget proposal, a general CRP sign-up could be held in FY2010, pending the publication of an environmental impact statement (EIS). Nationally there are 734,648 active contracts on 411,959 farms with 31.2 million acres enrolled, according to FSA's February 2010 monthly program summary.
Leading States	Leading states in terms of acres are TX (3.3 million), MT (3.1 million), and KS (2.8 million). Leading states in terms of number of contracts, are IA (102,759), IL (78,805), and MN (61,425).
Backlog/Interest	General sign-ups were not held in FY2007, FY2008, or FY2009. Contract extensions were offered to contracts that expired September 30, 2009. Of the 1.5 million acres that were offered extensions, 1.1 million acres were extended. Under continuous sign-ups, a total of 4.4 million acre were under contract as of the end of FY2009. About 400,000 acres are projected to be enrolled under continuous sign-up in FY2010.
Funding authority	Mandatory. No more than 32 million acres enrolled at any one time. No funding amount specified.
FY2010 funding (est.)	$1.96 billion (based on the estimated number of acres that will be enrolled and includes technical assistance).
FY2011 Administration request	$1.97 billion (based on the estimated number of acres that will be enrolled and includes technical assistance).

Statutory authority	Authorized in §1231-§1235 of the Food Security Act of 1985 (P.L. 99-198), as amended. Amended in §2101 of the Food, Conservation, and Energy Act of 2008 (P.L. 110-246). 16 U.S.C. 3831(a)-3835a.
Authorization expires	September 30, 2012.
Program website	http: // www. fsa .usda.gov/ FSA/webapp? area = home & subject = copr & topic = crp

CRP—Conservation Reserve Enhancement Program (CREP)

Administering Agency(s)	FSA, technical assistance by NRCS
Program Description	This subprogram of CRP partners with states at their request. States propose sub-state areas, such as a watershed, where environmental or resource concerns are more concentrated and can be addressed by enrolling up to 100,000 acres per project. States contribute 20% of the funding so that larger payments can be made, in order to encouraging greater participation.
Major 2008 farm bill amendments	None.
National Scope	34 states have active CREP contracts (67,498) on 44,570 farms, enrolling a total of 1.18 million acres, according to FSA's February 2010 monthly program summary.
Leading States	Leading states in terms of acres enrolled are PA (204,048), IL (126,631), and OH (103,093). States leading in number of contracts are OH (11,162), PA (10,961), and IL (6,624).

Backlog/Interest	Not applicable since any eligible land can be enrolled at any time; participation has been much higher in some states than in others, but that is due, reportedly, to how the program is promoted. Average rental payments are higher than for acreage under the general CRP sign up process.
Funding authority	Unspecified acreage subset of CRP.
FY2010 funding	Unspecified acreage subset of CRP.
FY2011 Administration request	Unspecified acreage subset of CRP.
Statutory authority	Authority derived from CRP statutory authority (see "Conservation Reserve Program (CRP)").
Authorization expires	September 30, 2012.
Program website	http:// www. fsa. usda.gov /FSA / webapp?area =home & subject = copr& topic=cep

CRP—Farmable Wetlands Program

Administering Agency(s)	FSA, technical assistance by NRCS
Program Description	This 1 million acre subprogram of the CRP enrolls small isolated agricultural wetlands. On a single tract of land, enrollment is now set at a maximum of 40 contiguous wetland acres. "Flooded farmland" has a 20-acre limit. Eligible lands include wetlands that were cropped 3 of the preceding 10 years, and buffers sufficient to protect them, on which the hydrology will be restored and a vegetative cover established.

Major 2008 farm bill amendments	Caps enrollment at 100,000 acres in any state and 1 million acres total. Expands enrollment of wetland and buffer acreage to include land that had been cropped during 3 of 10 crop years prior to 2002 and after 1990 and is subject to a natural overflow of a prairie wetland. Increases the maximum wetland size to 40 contiguous acres and makes all acres eligible for payment.
National Scope	Active contracts in 17 participating states, including: IL, IN, IA, KS, LA, MD, MI, MN, MS, MO, MT, NE, ND, OH, OK, SD, WI. According to FSA's February 2010 monthly program summary, 12,815 contracts on 10,165 farms have enrolled a total of 213,411 acres.
Leading States	In terms of acres, the leading states are IA (75,698 acres), SD (57,013 acres), and MN (41,972 acres). The largest number of contracts are in IA (4,824), followed by MN (3,065) and SD (3,035).
Backlog/Interest	Not applicable since any eligible land can be enrolled at any time; participation has been much higher in some states than in others, but that is due, reportedly, to how the program is promoted.
Funding authority	Mandatory. No more than 1 million acres enrolled at any one time and no more than 100,000 acres in any state (may be increased to 150,000 acres after three years).
FY2010 funding	Unspecified acreage subset of CRP.
FY2011 Administration request	Unspecified acreage subset of CRP.
Statutory authority	Authorized in Title XI of Agriculture and Related Agency appropriations, 2001 (P.L. 106-387)

	as §1231B of the Food Security Act of 1985 (P.L. 99-198), as amended. Amended by §2106 of the Food, Conservation, and Energy Act of 2008 (P.L. 110-246). 16 U.S.C. 3831b.
Authorization expires	September 30, 2012.
Program website	http://www.fsa.usda.gov /FSA / webapp? area = home& subject = copr & topic =fwp

Conservation Security Program

Administering Agency(s)	NRCS
Program Description	The 2008 farm bill terminated funding for new contracts under this program. The program was replaced by the "Conservation Stewardship Program (CSP)," discussed below. CSP provided financial and technical assistance for improvements in conserving environmental resources on farmland that meets soil and water quality criteria standards of NRCS. CSP supported producers who are already implementing and maintaining substantial conservation systems to protect soil, water, air, and wildlife, or who adopted more sustainable systems as part of the program. The maximum annual assistance through CSP increased with each of 3 progressive tiers of participation, each requiring greater conservation. Only producers in specified watersheds could participate each year.
Major 2008 farm bill amendments	The program was terminated and replaced with the "Conservation Stewardship Program (CSP)," discussed below. Payments on existing contracts are still authorized.

National Scope	Sign ups were held in FY2004, FY2005, FY2006, and FY2008 in designated watersheds. No new sign ups are allowed by law.
Leading States	For the FY2008 sign up, Ohio had the largest number of contracts (150), and Nebraska had the largest payments approved ($3,080,807). In FY2009, states with the most funds paid were MO ($29.5 million), OR ($24 million), and IA ($22.6 million).
Backlog/Interest	Not applicable.
Funding authority	Mandatory. Current law only allows payments on contracts entered into before September 30, 2008. No new contracts are allowed.
FY2010 funding (est.)	$233.9 million (payment on existing contracts only).
FY2011 Administration request	$212.4 million (payment on existing contracts only).
Statutory authority	Authorized in §2001 of the Farm Security and Rural Investment Act of 2002 (P.L. 107-171) as §1238-§1238C of the Food Security Act of 1985 (P.L. 99-198), as amended. Terminated in §2301 of the Food, Conservation, and Energy Act of 2008. 16 U.S.C. 3833-3838c.
Authorization expires	September 30, 2008 for new contracts. Payments are still authorized on contracts entered into before September 30, 2008.
Program website	http://www.nrcs.usda.gov/programs/csp/

Conservation Stewardship Program (CSP)

Administering Agency(s)	NRCS
Program Description	The new CSP replaces the Conservation Security Program and provides financial and technical assistance to promote the conservation and improvement of soil, water, air, energy, plant and animal life, and other conservation purposes on tribal and private working lands. Contracts (5 years in length with the option of extension) are based on meeting or exceeding a "stewardship threshold." Payments are based on the actual costs of installing conservation measures, any foregone income, and the value of the expected environmental outcomes.
Major 2008 farm bill amendments	The new CSP will continue to encourage conservation practices on working lands, but will be different from the former Conservation Security Program. It eliminates the three-tier approach, establishes 5-year rather than 10-year contracts, and requires direct attribution of payments, among other changes, thus requiring that USDA promulgate new rules for the program.
National Scope	The first CSP sign-up was held August 10, 2009. The first ranking period cut-off was September 30, 2009. During this period, over 21,200 applications were received, covering an estimated 33 million acres.
Leading States	During the FY2009 ranking period, NE had the most applications (2,683), followed by TX (1,879) and MO (1,821). TX had the most total acres apply (4.7 million), followed by NE (4.3 million) and NM (3.3 million). The most funding obligated in FY2009 was in TX ($766,575), MO ($482,455), and IA ($392,342).

Backlog/Interest	None identified..
Funding authority	Mandatory. 12.5 million acres each fiscal year. No funding amount specified.
FY2010 funding (est.)	$469.4 million (based on the estimated number of acres enrolled).
FY2011 Administration request	$629.5 million (based on the estimated number of acres enrolled).
Statutory authority	Authorized in §2301 of the Food, Conservation, and Energy Act of 2008 (P.L. 110-246) as §1238D-§1238G of the Food Security Act of 1985 (P.L. 99-198). 16 U.S.C. 3838h-3838n.
Authorization expires	September 30, 2012.
Program website	http://www.nrcs.usda.gov/ pograms/new_csp/csp.html

Emergency Conservation Program (ECP)

Administering Agency(s)	FSA, technical assistance by NRCS
Program Description	Provides emergency funding and technical assistance to producers to rehabilitate farmland damaged by natural disasters (e.g., hurricanes, floods, wind, and erosion) through activities such as removing debris, and implementing emergency water conservation measures in response to severe droughts.
Major 2008 farm bill amendments	None.

National Scope	Available nationwide. Participation varies widely and unpredictably from year to year. An FY2009 consolidated appropriations act (P.L. 110-329) provided $115 million to be available until expended. An FY2009 supplemental (P.L. 111-32) redesignated prior ECP funding for current use to be available until expended.
Leading States	Not applicable.
Backlog/Interest	Not applicable.
Funding authority	Discretionary. No funding amount specified.
FY2010 funding to date	$0
FY2011 Administration request	$0
Statutory authority	Authorized in §401 of the Agriculture Credit Act of 1978 (P.L. 95-334), as amended. 16 U.S.C. 2201-2205.
Authorization expires	Permanent authorization.
Program website	http://www.fsa.usda.gov/ FSA/webapp?area=home &subject=copr&topic=ecp

Emergency Watershed Program (EWP)

Administering Agency(s)	NRCS on private lands, U.S. Forest Service on National Forest Systems lands

Program Description	EWP provides technical and financial assistance to reduce hazards to life and property in watersheds that have been damaged by natural disasters. Assistance includes disaster cleanup and recovery activities, and purchasing easements in flood plains that will benefit natural resources such as wetlands, while reducing the risk of exposure to future natural disasters.
Major 2008 farm bill amendments	None.
National Scope	Available nationwide. Participation varies widely and unpredictably from year to year. In FY2008 and FY2009, Congress provided two supplemental appropriations totaling almost $500 million for hurricane, flood, and fire recovery. An additional $145 million was provided for floodplain easements through the American Recovery and Reinvestment Act of 2009.
Leading States	Not applicable.
Backlog/Interest	Not applicable.
Funding authority	Discretionary. No funding amount specified.
FY2010 funding to date	$0
FY2011 Administration request	$0
Statutory authority	Authorized in §216 of P.L. 81-516 and §403 of the Agriculture Credit Act of 1978 (P.L. 95- 334), as amended. 16 U.S.C. 2203; and 33 U.S.C. 701b-1.
Authorization expires	Permanent authorization.

Program website	Recovery: http://www.nrcs.usda.gov/programs/ewp/; Floodplain easements: http://www.nrcs.usda.gov/programs/ewp/Floodplain/index.html

Environmental Quality Incentives Program (EQIP)

Administering Agency(s)	NRCS
Program Description	EQIP provides financial and technical assistance to producers and land owners to plan and install structural, vegetative, and land management practices on eligible lands to alleviate natural resource problems. Eligible producers enter into contracts to receive payment for implementing conservation practices. Approved activities are carried out according to an EQIP plan developed in conjunction with the producer that identifies the appropriate conservation practice(s) to address resource concerns on the land. Sixty percent of the funds are targeted to livestock producers. EQIP is the largest working lands program.
Major 2008 farm bill amendments	Lowers the limitation on payments to a total of \$300,000, down from \$450,000; limits projects with organic production benefits to \$20,000 annually or \$80,000 in any six-year period; authorizes a new water enhancement program (see "EQIP— Agricultural Water Enhancement Program (AWEP)"); reauthorizes the innovative grants program (see "EQIP— Conservation Innovation Grants (CIG)"); retains the allocation of sixty percent of funding each year to practices related to livestock production.
National Scope	Available nationwide. In FY2009, EQIP allocated \$1.1 billion for 31,960 contracts covering 12

	million acres, according to information in the FY2011 budget notes.
Leading States	In FY2009, the top three state by contracts signed were TX (3,063), CA (1,743) and MS (1,684). The most funding obligated was in TX ($83.1 million), CA ($69.9 million), and MN ($34.8 million).
Backlog/Interest	In FY2009, 31,960 applications were funded and 54,329 applications went unfunded. Only 37% of valid applications were funded in FY2009, according to information in the FY2011 budget notes. The total estimated cost of this backlog is $1.36 billion. The most unfunded applications were submitted in OK (4,063), TX (3,944), and CA (3,853).
Funding authority	Mandatory. FY2009 - $1.377 billion, FY2010 - $1.45 billion, FY2011 - $1.588 billion, FY2012 - $1.75 billion.
FY2010 funding	$1.18 billion ($1.45 billion authorized).
FY2011 Administration request	$1.21 billion ($1.59 billion authorized).
Statutory authority	Authorized in subtitle D of Title III (§331-336) of the Federal Agriculture Improvement and Reform Act of 1996 (P.L. 104-127) as §1240-§1240I of the Food Security Act of 1985 (P.L. 99-198), as amended. Amended by §2501 of the Food, Conservation, and Energy Act of 2008 (P.L. 110-246). 16 U.S.C. 3839aa-3839aa90.
Authorization expires	September 30, 2012
Program website	http://www.nrcs.usda.gov/programs/eqip/

EQIP—Agricultural Water Enhancement Program (AWEP)

Administering Agency(s)	NRCS
Program Description	AWEP is a subprogram of EQIP. The program provides financial and technical assistance to producers and land owners to address water quality and quantity concerns on agricultural land. The program replaces the Ground and Surface Water Conservation program and the Klamath Basin program. AWEP addresses water quality and quantity concerns through eligible partners and established priority areas.
Major 2008 farm bill amendments	Program created in the 2008 farm bill.
National Scope	Available nationwide, although not every state participated in FY2009. In FY2009, AWEP allocated $58 million to 1,687 contracts covering 488,380 acres, according to information in the FY2011 budget notes. A total of 63 project area proposals were approved in the national competition. Of these, 6 projects were with tribes, 2 were multi-state, and 24 were in a special priority area.
Leading States	In FY2009, the top three states by contracts signed were CA (555), GA (263), and TX (225). The most funding obligated was in CA ($20.8 million), TX ($12.7 million), and ID ($7.9 million). States with the most total applications were CA (890), NE (703), and GA (576).
Backlog/Interest	In FY2009, 1,687 contracts were funded and 1,155 applications went unfunded. The total estimated cost of this backlog is $40.6 million. The highest numbers of unfunded applications were in NE (412), GA (284), and AL (115).

Funding authority	Mandatory. FY2009 - $73 million, FY2010 - $73 million, FY2011 - $74 million, FY2012 and each year thereafter - $60 million.
FY2010 funding	$73 million.
FY2011 Administration request	$74 million.
Statutory authority	Authorized in subtitle F of Title II (§2510) of the Food, Conservation, and Energy Act of 2008 (P.L. 110-246) as §1240I of the 1985 Food Security Act (P.L. 99-198). 16 U.S.C. 3839aa-9.
Authorization expires	Permanent authorization.
Program website	http://www.nrcs.usda.gov/programs/awep/

EQIP—Conservation Innovation Grants (CIG)

Administering Agency(s)	NRCS
Program Description	CIG is a subprogram of EQIP that awards competitive grants to state and local agencies, non-governmental organizations, tribes, and individuals to implement innovative conservation techniques and practices. The program was first implemented in FY2004. Annual requests for proposals are published in the *Federal Register* and include separate funding categories for national, state, technology, and Chesapeake Bay competitions. Examples of eligible projects include market systems for pollution reduction, demonstrating precision agriculture,

	capturing nutrients through a community anaerobic digester, and establishing a tribal partnership for regional habitat conservation.
Major 2008 farm bill amendments	Amends the grants to cover air quality concerns associated with agriculture (including greenhouse gas emissions); and expands the CIG objectives to include forest resource management and projects that provide conservation benefits through increased participation by producers of specialty crops.
National Scope	Available nationwide with select states offering state competitions. CIG awarded a total of $18.4 million (55 projects) in FY2009.
Leading States	Not applicable.
Backlog/Interest	None identified.
Funding authority	Unspecified subset of EQIP.
FY2010 funding	Unspecified subset of EQIP. For the past five years, approximately $20 million of EQIP's authorized funding has been available for CIG.
FY2011 Administration request	Unspecified subset of EQIP.
Statutory authority	Authorized in §2301 of the Farm Security and Rural Investment Act of 2002 (P.L. 107-171) as §1240H of the Food Security Act of 1985 (P.L. 99-198), as amended. Amended by §2509 of the Food, Conservation, and Energy Act of 2008 (P.L. 110-246). 16 U.S.C. 3839aa-8.
Authorization expires	September 30, 2012.

Program website	http://www.nrcs.usda.gov/programs/cig/

Farmland Protection Program (FPP)[4]

Administering Agency(s)	NRCS
Program Description	FPP provides funds to state, tribal, and local governments, and non-governmental organizations to help them purchase conservation easements from willing sellers to limit conversion of farmland to nonagricultural uses. The length of time from when a land owner first offers to sell an easement to when it is recorded with the deed to the land can be considerable.
Major 2008 farm bill amendments	Changes the program's purpose from protecting topsoil to protecting the land's agricultural use by limiting nonagricultural uses. The program is also restructured to emphasize longer term and renewable cooperative agreements.
National Scope	Available nationwide; however, some states may not have existing farmland protection programs through which FPP is carried out. From the program's inception in FY1996 through FY2009, $716 million was spent to acquire 2,380 easements on 456,600 acres, according to the FY2011 budget notes.
Leading States	In FY2009, the largest amount was obligated in NJ ($6.9 million), followed by MA ($6.1 million) and WA ($5.9 million).
Backlog/Interest	The FY2011 budget notes state that "The demand for the program has exceeded available funds by approximately 200 percent."

Funding authority	Mandatory. FY2009 - $121 million, FY2010 - $150 million, FY2011 - $175 million, and FY2012 - $200 million.
FY2010 funding	$150 million.
FY2011 Administration request	$160 million ($175 million authorized).
Statutory authority	Authorized in §388 of the Federal Agriculture Improvement and Reform Act of 1996 as §1238H-§1238I of the 1985 Food Security Act (P.L. 99-198), as amended. Amended by §2401 of the Food, Conservation, and Energy Act of 2008 (P.L. 110-246). 16 U.S.C. 3838h-3838i.
Authorization expires	September 30, 2012.
Program website	http://www.nrcs.usda.gov/programs/frpp/

Grassland Reserve Program (GRP)

Administering Agency(s)	Easements are administered by NRCS, agreements are administered by FSA, and NRCS provides technical assistance for both.
Program Description	GRP uses long-term rental agreements and easements to help land owners and producers restore and protect grasslands while maintaining them in a condition suitable for grazing using common management practices.
Major 2008 farm bill amendments	Directs 40% of funds to rental contracts (10-, 15-, and 20-year duration) and 60% for permanent easements; authorizes the enrollment of 1.22 million acres between FY2009 and FY2012; gives priority

	enrollment to certain expiring CRP lands; tribal lands are now eligible for the program.
National Scope	Available nationwide. In FY2009, a total of $40.5 million was obligated and committed, $24.3 million to GRP easements and $16.2 million to rental contracts. In FY2009, 511 new participants enrolled 146,269 acres, including 56,688 permanent easements, according to information in the FY2011 budget notes.
Leading States	In FY2009, states with the most funding obligated were TX ($333,475), NM ($232,312), and OK ($152,475).
Backlog/Interest	None identified.
Funding authority	Mandatory. Authorized to enroll 1.22 million acres between FY2009-FY2012.
FY2010 funding (est.)	$100.7 million (based on the estimated number of acres that will be enrolled).
FY2011 Administration request	$79.1 million (based on the estimated number of acres that will be enrolled).
Statutory authority	Authorized in §2401 of the of the Farm Security and Rural Investment Act of 2002 (P.L. 107-171) as §1238N-§1238Q of the Food Security Act of 1985 (P.L. 99-198), as amended. Amended by §2403 of the Food, Conservation, and Energy Act of 2008 (P.L. 110-246). 16 U.S.C. 3838n-3838q.
Authorization expires	September 30, 2012.
Program website	http://www.nrcs.usda.gov/programs/GRP/

Healthy Forests Reserve Program (HFRP)

Administering Agency(s)	NRCS
Program Description	Assists landowners in restoring and enhancing forest ecosystems using 10-year agreements, 30-year easements, and permanent easements.
Major 2008 farm bill amendments	Allows permanent easements; directs 40% of funds to cost-share agreements and 60% for 30- year or permanent easements; and authorizes mandatory funding through the CCC.
National Scope	Not available nationwide. Limited participation in AR, MS, OK, GA, IN, MN, ME, and OR.
Leading States	The program was initially introduced as a pilot program in AR, ME, and MS. In FY2009, nine applications were enrolled into easements in MS and AR. Eight are 30-year easements and one is a permanent easement. .
Backlog/Interest	During the FY2009 sign-up, a total of 122 applications covering 19,426 acres were submitted. Nine easements were enrolled in two states, covering a total of 809 acres.
Funding authority	Mandatory. $9.75 million annually for FY2009-FY2012.
FY2010 funding	$9.75 million.
FY2011 Administration request	$9.75 million.

Statutory authority	Authorized in Title V Healthy Forest Restoration Act of 2003 (P.L. 108-148), as amended. Amended by §8205 of the Food, Conservation, and Energy Act of 2008 (P.L. 110-246). 16 U.S.C. 6572.
Authorization expires	September 30, 2012.
Program website	http://www.nrcs.usda.gov/ programs/HFRP/Prog Info/Index.html

Resource Conservation and Development (RC&D) Program

Administering Agency(s)	NRCS
Program Description	The RC&D program provides support in the form of NRCS staff coordinators to authorized county and multi-county areas. Coordinators assist state and local units of government and non-profits to develop and carry out programs to conserve and improve natural resources and the use of land, and improve conditions in rural America.
Major 2008 farm bill amendments	Emphasizes the locally led planning process to provide assistance for implementing area plans.
National Scope	375 authorized areas encompass approximately 2,696 counties, more than 85% of the national total, and more than 77% of the U.S. population, according to the FY2011 budget notes.
Leading States	Not applicable.
Backlog/Interest	38 application areas covering 231 additional counties have pending applications for the Secretary's designation.
Funding authority	Discretionary. No amount specified.

FY2010 funding	$50.7 million.
FY2011 Administration request	$0
Statutory authority	Authorized in §31 and §32 of the Bankhead-Jones Farm Tenant Act (P.L. 89-796), as amended. Amended by §2805 of the Food, Conservation, and Energy Act of 2008 (P.L. 110- 246). 16 U.S.C. 3451-3459.
Authorization expires	Permanent authorization.
Program website	http://www.nrcs.usda.gov/programs/rcd/

Watershed and Flood Prevention Operations

Administering Agency(s)	NRCS
Program Description	Also referred to as the Small Watershed Program, Watershed and Flood Prevention Operations are two separate authorizations under which more than 11,000 structures have been built in more than 1,500 active and completed projects. The P.L. 78-534 Flood Prevention Operations Program authorizes 11 projects, while the P.L. 83-566 Small Watershed Operations Program authorizes watershed projects generally. Projects may be authorized for any of 8 purposes; almost all projects have flood control as an authorized purpose. Under P.L. 566, NRCS provides technical and financial assistance to plan and install projects on private lands, in cooperation with local sponsors, states, and other public agencies. The small watershed project costs are shared with local partners. Projects are limited to a maximum size,

	including 25,000 acre-feet of total capacity and 250,000 acres in extent. Projects above a specified size require congressional committee authorization.
Major 2008 farm bill amendments	None.
National Scope	Available nationwide. A total of 397 work plans are complete under P.L. 534, and 1,360 are active or completed under P.L. 566, according to the FY2011 budget notes. Six new P.L. 566 projects were authorized for FY2009, the first new authorized projects since FY2005.
Leading States	In FY2009 the state with the most obligations was MO ($43.1 million), followed by FL ($29.9 million) and LA ($28.2 million).
Backlog/Interest	NRCS identifies a total of $1.25 billion in unfunded federal commitments for approximately 300 authorized projects; the greatest value of unfunded commitments is in OK ($271 million) and TX ($245 million), according to the FY2011 budget notes.
Funding authority	Discretionary. No amount specified.
FY2010 funding	$30 million.
FY2011 Administration request	$0
Statutory authority	Authorized in the Flood Control Act of 1944 (P.L. 78-534), as amended, and the Watershed Protection and Flood Prevention Act (P.L. 83-566), as amended. 33 U.S.C. 701b-1 and 16 U.S.C. 1000, *et. seq*.
Authorization expires	Permanent authorization.

Program website	http://www.nrcs.usda.gov/programs/watershed/index.html

Watershed Rehabilitation Program

Administering Agency(s)	NRCS
Program Description	The Watershed Rehabilitation Program provides technical and financial assistance for planning, design, and implementation to rehabilitate aging watershed dam projects (including upgrading or removing dams) in communities to address health and safety concerns. Only dams constructed under the Watershed and Flood Prevention Operations program (see "Watershed and Flood Prevention Operations" above) are eligible. Small watershed project dams have a 50-year design life, and 1,344 reached or exceeded that time span by the end of 2008. By the end of 2015, this number will exceed 4,300, according to the FY2011 budget notes.
Major 2008 farm bill amendments	Reauthorizes both mandatory and discretionary funding.
National Scope	Only available for dams built through the Watershed and Flood Prevention Operations program and the RC&D program (see discussions above). In FY2009, a total of 82 projects were funded in 21 states, and 650 dam assessments were funded in 27 states, according to the FY2011 budget notes.
Leading States	The largest number of funded dam projects (2000-2009) are in OK (44), MS (20), and TX (17); States with the most obligations in FY2009 were AZ ($7.2 million), OK ($6.8 million), and TX ($4.2 million).
Backlog/Interest	In FY2009, over $14 million in requests were received from public sponsors of 716 dams to

	have the condition of their dams assessed in consideration for rehabilitation.
Funding authority	Mandatory: FY2009, $100 million. Discretionary: $65 million annually, FY2009-FY2012.
FY2010 funding	$40.2 million in discretionary funding and $0 in mandatory funding.
FY2011 Administration request	$40.5 million in discretionary funding and $0 in mandatory funding.
Statutory authority	Authorized in §313 of the Grain Standards and Warehouse Improvement Act of 2000 (P.L. 106-472) as §14 of the Watershed Protection and Flood Prevention Act, as amended. Amended by §2803 of the Food, Conservation, and Energy Act of 2008 (P.L. 110-246). 16 U.S.C. 1012
Authorization expires	September 30, 2012.
Program website	http://www.nrcs.usda.gov/programs/WSRehab/

Wetlands Reserve Program (WRP)

Administering Agency(s)	NRCS
Program Description	WRP funds the purchase of easements (30 years or permanent), restoration agreements (usually 10 years in length), and 30-year contracts to assist land owners in protecting and restoring wetlands. It provides technical and financial assistance, and emphasizes restoration to original natural conditions where possible.

Major 2008 farm bill amendments	Increases the WRP maximum enrollment cap to over 3.014 million acres (up from a previous cap of 2.275 million acres); expands eligible lands to include certain types of private and tribal wetlands, croplands, and grasslands, as well as lands that meet the habitat needs of specific wildlife species; changes the payment schedule for easements; limits wetland restoration payments; specifies criteria for ranking program applications; requires USDA conduct an annual survey of the Prairie Pothole Region starting with FY2008; and requires USDA to submit a report to Congress on long-term conservation easements under the program.
National Scope	Available nationwide. Through FY2009, 2.18 million acres have been enrolled, with easements closed and restored on 1.7 million acres. In 2009, more than 179,000 acres were enrolled in over 1,000 projects, according to the FY2011 budget notes.
Leading States	In FY2009, the top three states with the most funding obligated were FL ($77.5 million), MN ($31.5 million), and WI ($25.5 million).
Backlog/Interest	Not available.
Funding authority	Mandatory. Maximum enrollment of 3.014 million acres at any one time. No funding amount specified.
FY2010 funding (est.)	$613.1 million (based on the estimated number of acres that will be enrolled).
FY2011 Administration request (est.)	$502.2 million (based on the estimated number of acres that will be enrolled).
Statutory authority	Authorized in §1438 of the Food, Agriculture, Conservation and Trade Act of 1990 (P.L. 101-

	624) as §1237-§1237F of the 1985 Food Security Act (P.L. 99-198), as amended. Amended by §2201-2208 of the Food, Conservation, and Energy Act of 2008 (P.L. 110-246). (16 U.S.C. 3837-3837f)
Authorization expires	September 30, 2012.
Program website	http://www.nrcs.usda.gov/programs/wrp/

Wildlife Habitat Incentive Program (WHIP)

Administering Agency(s)	NRCS
Program Description	WHIP provides technical and financial assistance to eligible participants to develop upland wildlife, wetland wildlife, threatened and endangered species, fish and other types of wildlife habitat. Participants enter into contracts, usually 5 to 10 years in duration, which pays up to 75% of the contract implementation cost.
Major 2008 farm bill amendments	Limits program eligibility to focus on "the development of wildlife habitat on private agricultural land, nonindustrial private forest land, and tribal lands," thus potentially excluding some previously covered areas (i.e., non-agricultural lands). It also allows USDA to provide priority to projects that address issues raised by state, regional, and national conservation initiatives.
National Scope	Available nationwide. In FY2009, over 3,700 agreements were enrolled on over 800,000 acres.
Leading States	In FY2009, the largest number of contracts was in MS (298), followed by MN (207) and AL (142). The largest number of enrolled acres was in AZ (176,251 acres), NM (132,008 acres), and TX (78,558 acres).

	The greatest obligation of funding was in TX ($3.96 million), followed by AK ($2.08 million) and CA ($1.87 million).
Backlog/Interest	In FY2009, WHIP had 2,207 unfunded applications at a total cost of $34.8 million. OK had the most unfunded applications (332), followed by VT (155) and WV (149). OK had the highest cost of these applications ($5.1 million), followed by WV ($2.2 million) and ND ($2.1 million).
Funding authority	Mandatory. $85 million annually FY2009-FY2012.
FY2010 funding	$85 million.
FY2011 Administration request	$73 million ($85 million authorized).
Statutory authority	Authorized in §387 of the Federal Agriculture Improvement and Reform Act of 1996 (P.L. 104-127) as §1240N of the 1985 Food Security Act (P.L. 99-198), as amended. Amended by §2602 of the Food, Conservation, and Energy Act of 2008 (P.L. 110-246). (16 U.S.C. 3839bb- 1)
Authorization expires	September 30, 2012.
Program website	http://www.nrcs.usda.gov/programs/whip/

Author Contact Information

Megan Stubbs
Analyst in Agricultural Conservation and Natural Resources Policy
mstubbs@crs.loc.gov, 7-8707

End Notes

[1] Conservation and Trade Act of 1990 (P.L. 101-624), Federal Agricultural Improvement and Reform Act of 1996 (P.L. 104-127), Farm Security and Rural Investment Act of 2002 (P.L. 107-171), and Food, Conservation, and Energy Act of 2008 (P.L. 110-246).

[2] The CCC is the funding mechanism for the mandatory payments that are administered by various agencies of USDA, including all of the farm commodity price and income support programs.

[3] For more information on: ARS projects, see http://www.ars.usda.gov/Research/Research.htm; ERS projects, see http://www.ers.usda.gov/Browse/view. aspx?subject= Natural Resources Environment; and FS projects, see http://www.fs.fed.us/projects/.

[4] USDA calls this program the Farm and Ranch Lands Protection Program (FRPP).

In: Agricultural Conservation: Programs and Issues ISBN: 978-1-61761-288-6
Editors: Steven N. Murphy

Chapter 2

AGRICULTURAL CONSERVATION ISSUES IN THE 111TH CONGRESS

Megan Stubbs

SUMMARY

Agricultural conservation has been a public policy issue for more than 60 years. Congress has repeatedly taken action on the issue through water and soil legislation, often as part of omnibus farm bills. Early policy decisions were directed at addressing natural resource concerns on the farm, primarily reducing high levels of soil erosion and providing water to agriculture in quantities and quality that enhanced farm production. In more recent years, conservation policy has shifted to concerns about the off-farm impacts of agricultural activities.

The latest farm bill, the Food, Conservation, and Energy Act of 2008 (P.L. 110-246), reauthorized most existing conservation programs, modified several programs, and created a few new programs. The U.S. Department of Agriculture's (USDA's) conservation efforts have centered on implementing these conservation programs through working land conservation practices, retiring land from production or establishing conservation easements, and providing technical assistance. Program implementation controversies could lead to congressional oversight or action, especially given the recent financial statement audit reports on conservation program payments at USDA.

Other emerging issues in the 111th Congress could have a significant impact on agricultural conservation. The climate change debate and use of ecosystem services markets has brought conservation to the forefront of discussion on the role of agriculture in reducing greenhouse gas emissions. Also, the effect of ethanol production on natural resources and changes in land use is an ongoing concern in the area of biofuels policy. Other environmental issues for agriculture such as concentrated animal feeding operation regulations, greenhouse gas emission reporting for livestock producers, and wetlands mitigation could lead to expanded opportunities and challenges for many conservation efforts.

Appropriations and budget issues continue to influence conservation programs and policy. Conservation programs with mandatory funding have been routinely reduced through annual appropriation bills. In FY2009 and FY2010, the reductions were limited to a handful of programs, with the Environmental Quality Incentives Program (EQIP) receiving the largest reductions, including $270 million in FY2010. The watershed programs have experienced an increase in congressionally directed projects through appropriations, with Watershed and Flood Prevention Operations being 97% earmarked in FY2009 and 74% in FY2010. The ongoing issue of funding for conservation technical assistance in mandatory programs will likely be raised again due to an expiring authority.

Other issues of potential interest in the 111th Congress include implementation of conservation program payment and income limitations, use of the Conservation Effects Assessment Project, and recent financial audits and conservation contract administration concerns.

INTRODUCTION

Agricultural conservation has been a public policy issue for more than 60 years, an issue repeatedly recast as new problems have emerged or old problems have resurfaced. Early conservation efforts were focused on two themes—reducing high levels of soil erosion and providing water to agriculture in quantities and quality that enhanced farm production.

Congress responded repeatedly to these themes by creating or revising programs designed to reduce resource problems on the farm. By the early 1980s, however, concern was growing that these programs were not adequately dealing with environmental problems resulting from agricultural activities (especially off the farm). Publicized instances of significant problems, especially high soil erosion rates said to rival the dust bowl era, increased awareness and intensified the policy debate. In 1985, conservation policy took a new direction when Congress enacted four major new conservation programs in the conservation title of the Food Security Act of 1985 (P.L. 99-198): the Conservation Reserve Program (CRP), Sodbuster, Conservation Compliance, and Swampbuster. These programs, all administered by the U.S. Department of Agriculture (USDA), were the first conservation programs designed to deal with environmental problems related to the off-farm impacts of agricultural activities.

Provisions enacted in subsequent farm bills, including in 1990, 1996, 2002, and 2008, reflect a rapid evolution of the conservation agenda, including the growing influence of environmentalists and other non-agricultural interests in the formulation of conservation policy, and a recognition that agriculture was not treated like other business sectors in many environmental laws.[1] Congress also began funding many of these new programs through mandatory spending for the first time, using the borrowing authority of USDA's Commodity Credit Corporation (CCC)[2] as the funding mechanism instead of annual appropriations. In addition to the original soil erosion and water quality and quantity issues, the conservation agenda has continued to expand to address other natural resource concerns such as wildlife habitat, air quality, wetlands restoration and protection, energy efficiency, and sustainable agriculture.

Many conservation programs receive continued support among producers and Congress, as evidenced by the recent reauthorization and expansion of existing programs and creation of new ones in the most recent farm bill (Food, Conservation, and Energy Act of 2008, P.L. 110-246). The enacted bill addressed several emerging issues, such as payment structure, geographic targeting, program complexity, the importance of large-scale conservation efforts, and measurement of costs and effectiveness. Funding also played a large role in the 2008 farm bill debate as a pronounced domestic budget constraint caused several conservation groups to compete with other farming interests for the necessary resources to expand and continue many conservation programs. Many of these general policy issues were central to outcomes in the enacted conservation title of the farm bill.

Current Major Activities

USDA's conservation efforts have centered in recent years on implementing conservation programs that target working lands, remove land from production (also known as retiring land) or create conservation easements, and provide technical assistance. Funding for the overall conservation effort has grown significantly since the 2002 farm bill (P.L. 107-171) and many reauthorized programs received increases in funding in the 2008 farm bill. Recent trends in policy for the suite of conservation programs include less emphasis on land retirement and on land producing row crops, and more attention to conservation on land in other agricultural uses and to livestock producers. Congress in the 2008 farm bill expanded many program benefits to include specialty crop producers, producers transitioning to organic production, forested and managed lands, pollinator habitat and protection, and nutrient and pest management.

Lead agricultural conservation agencies within USDA are the Natural Resources Conservation Service (NRCS), which provides technical assistance and administers most programs, and the Farm Service Agency (FSA), which administers the largest program, CRP. These agencies are supported by others in USDA that supply research and educational assistance, including the Agricultural Research Service, the Forest Service, and the Economic Research Service.[3] In addition, the conservation effort involves a very large array of partners, including other federal agencies, state and local governments, and private organizations, among others, who provide funds, expertise, and other forms of assistance to the conservation effort.

Overview of Conservation Programs

There are currently more than 20 USDA agricultural conservation programs that assist private landowners with natural resource concerns. Conservation programs are grouped into working land programs, land retirement and easement programs, watershed programs, and other programs. For a brief description of the individual USDA agricultural conservation programs, see CRS Report R40763, *Agricultural Conservation: A Guide to Programs*, and for program funding levels, see the **Appendix**.[4] **Figure 1** illustrates the estimated amount of funding in FY2009 for each group of programs.

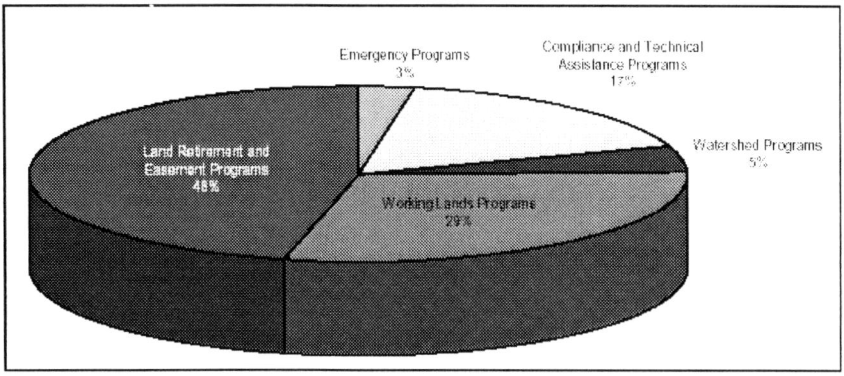

Source: CRS Funding estimates compiled from FY2009 Omnibus Appropriations Act (P.L. 111-8), American Recovery and Reinvestment Act of 2009 (P.L. 111-5), and USDA, FSA, *CCC Commodity Estimates Book*, FY2010 Mid-Session Review, August 25, 2009, http://www.fsa.usda.gov/FSA/webapp? area=about&subject = landing& topic= bap-bu-ce.

Notes: Technical assistance includes funding for compliance programs. Funding for extension and research is not included. Total funding for conservation programs is approximately $5.3 billion.

Figure 1. 2009 Estimated Funding for Conservation Programs

Working Lands

Working lands conservation programs are typically classified as programs that allow private land to remain in production, while implementing various conservation practices to address natural resource concerns specific to the area. The largest of these programs is the Environmental Quality Incentives Program (EQIP), currently authorized at a total of $7.3 billion between FY2008 and FY2012. Others, such as the Wildlife Habitat Incentives Program (WHIP), Agricultural Management Assistance (AMA), and Agricultural Water Enhancement Program (AWEP), operate similarly to EQIP; however, they target specific resource concerns or geographic areas. The new Conservation Stewardship Program (CSP) replaces the Conservation Security Program and is designed to encourage producers to address specific resource concerns in a comprehensive manner.

Land Retirement and Easement Programs

Land retirement programs provide federal payments to private agricultural landowners for *temporary* changes in land use or management to achieve environmental benefits. Conversely, conservation easements impose a *permanent* land-use restriction that is voluntarily placed on the land in exchange for a government payment. The largest land retirement program is the Conservation Reserve Program (CRP), which removes land from production and is authorized to enroll up to 32 million acres. Other programs such as the Wetlands Reserve Program (WRP) and the Grasslands Reserve Program (GRP) use a combination of long-term or permanent easements as well as restoration contracts to protect wetlands and grasslands from production. The Farmland Protection Program (FPP) also uses easements; unlike the aforementioned programs, however, it does not remove land from production, but rather restricts productive farmland from being developed for non-farm purposes.

Watershed Programs

Watershed and Flood Prevention Operations (P.L. 83-566 and P.L. 78-534) are two separate program authorities that have almost identical objectives and operations. These programs are sometimes referred to as the Small Watershed Programs. Through these programs, NRCS partners with local sponsors to carry out activities for soil conservation; flood prevention; conservation, development, utilization and disposal of water; and conservation and proper utilization of land. The Watershed Rehabilitation program was authorized in 2000 (P.L. 106-472) to rehabilitate watershed structures created under the Watershed and Flood Prevention Operations programs that have reached or exceeded their design life.

Other Selected Programs

USDA administers several other conservation programs. Technical assistance programs, such as Conservation Operations (including Conservation Technical Assistance, Survey, Soil Survey, Grazing Lands Conservation Initiative, and the Plant Materials Centers) provide landowners with science-based conservation information and technical expertise (e.g., engineering and biological) unique to the region and land-use type. USDA also administers compliance programs such as Sodbuster, Swampbuster, conservation compliance, and Sodsaver, which halt a producer's access to many federal farm program benefits if they do not meet conservation program requirements for highly erodible lands and wetlands. Programs such as the Emergency Conservation Program (ECP) and the Emergency Watershed Program (EWP) provide disaster assistance for farmland rehabilitation and impairments to watersheds, and are funded on an ad-hoc emergency basis.

USDA AGRICULTURAL CONSERVATION PROGRAMS

Working Lands Programs—Environmental Quality Incentives Program (EQIP), Conservation Stewardship Program (CSP)/Conservation Security Program, Wildlife Habitat Incentives Program (WHIP), Agricultural Water Enhancement Program (AWEP), and Agricultural Management Assistance (AMA).

Land Retirement and Easement Programs—Conservation Reserve Program (CRP, including Conservation Reserve Enhancement Program (CREP) and CRP Farmable Wetlands), Wetlands Reserve Program (WRP), Farmland Protection Program (FPP), Grassland Reserve Program (GRP), and Healthy Forests Reserve Program (HFRP).

Watershed Programs—Watershed and Flood Prevention Operations (also referred to as the Small Watershed Programs), Watershed Rehabilitation, and Watershed Surveys and Planning.

Emergency Programs—Emergency Conservation Program (ECP) and Emergency Watershed Program (EWP) (includes floodplain easements).

Compliance—Conservation Compliance, Sodbuster, Swampbuster, and Sodsaver.

Technical Assistance Programs—Conservation Operations (includes Conservation Technical Assistance, Survey, Soil Survey, Grazing Lands Conservation Initiative, Plant Materials Centers), Resource Conservation and Development (RC&D) program.

Other Conservation Programs and Provisions—Chesapeake Bay Watershed Program, Cooperative Conservation Partnership Initiative, Conservation Innovation Grants, Great Lakes Basin Program, Regional Equity, State Technical Committees, Voluntary Public Access and Habitat Incentive Program, and Grassroots Source Water Protection Program.

Agricultural Conservation Issues in the 111th Congress

2008 Farm Bill Implementation

The Food, Conservation, and Energy Act of 2008 (P.L. 110-246), the 2008 farm bill, reauthorizes almost all existing conservation programs, modifies several programs, and creates various new conservation programs. The changes address eligibility requirements, program definitions, enrollment and payment limits, contract terms, evaluation and application ranking criteria, among other administrative issues.[5] Following the passage of the 2008 farm bill, USDA has taken steps toward implementation of many conservation programs. In the 111th Congress, lawmakers' attention likely will be focused on USDA's implementation of these provisions. The following sections provide a brief description of implementation issues within some larger conservation programs and provisions.

Conservation Reserve Program (CRP)

The 2008 farm bill reauthorized CRP and reduced the maximum acreage cap to 32 million acres, down from a cap of 39.2 million acres established in the 2002 farm bill. With the reduced cap, USDA is less likely to offer reenrollment opportunities, except on the most environmentally beneficial cases with the highest environmental benefit index (known as EBI score) or highest potential to erode (known as the erodibility index, EI). **Figure 2** illustrates expiring CRP acreage eligible for extension because of high EBI and EI rankings. Many of these acres are located in the central region of the country within the central migratory bird flyway. Some wildlife organizations have expressed concern over this location due to the potential loss of benefits CRP acres provide to wildlife and migratory birds. Other conservation organizations see the acreage cap reduction as a result of CRP participants not reenrolling in the program as their 10-year contracts come up for renewal. They contend that the reduction is more realistic and allowed Congress to use the savings generated by the reduced cap to expand other conservation programs.

USDA announced a sign-up for voluntary extensions beginning May 18, 2009, and running through June 30, 2009. Of the expiring 3.8 million acres, approximately 180,000 acres were reenrolled under a continuous sign-up contract. Only about 1.5 million acres enrolled under the general sign-up were offered contract extensions, and approximately 70% or 1.1 million accepted a three- to five-year extension. Acreage eligible for contract extensions must score in the top 30% of the EBI or have an EI of 15 or higher. The reduction in contract extensions was necessary to ensure that the total acreage under contract would conform to the 32 million acre limit on October 1, 2009. As of October 31, 2009, FSA reported a total enrollment of 31.1 million acres following the expiration of 2.7 million acres. Texas, Colorado, Kansas, North Dakota, and South Dakota had the largest number of acres expire. These five states account for 60% of the total national acres that expired on October 1, 2009.[6]

Environmental Quality Incentives Program (EQIP)

With the 111th Congress experiencing continued budget constraints, EQIP could face similar challenges, with a potential reduction in mandatory funding levels and a continuing backlog of unfunded applications. The 2008 farm bill continued an increase in authorized mandatory funding that began with the 2002 farm bill. Mandatory funding for EQIP has grown substantially from its FY2002 level of $200 million, to the authorized level of $1.75 billion in FY2012. Despite this significant increase in authorized mandatory funding, annual appropriations acts have reduced the actual funding levels by a total of nearly $1.16 billion from FY2005 through FY2010.[7] Some of this reduction has come at the request of both the current and previous Administrations. The number of pending applications continues to exceed the amount of available funding (**Table 1**). In 2008, 67% of eligible applications were funded. Many conservation groups are concerned that this has deterred producers from applying and enrolling in the program. Others point out that despite reductions from the authorized level, total funding continues to increase. This issue will likely intensify if annual appropriations continue to reduce actual funding for EQIP below authorized mandatory levels.

Conservation Stewardship Program (CSP)

The original Conservation Security Program, established in the 2002 farm bill (P.L. 107-171), faced numerous implementation challenges, drawing criticism by many in Congress.[8] The 2008 farm bill established the new Conservation Stewardship Program (CSP) to restructure and replace the old CSP. On July 29, 2009, NRCS implemented CSP with an interim final rule published in the *Federal Register*.[9] Applications for the 2009 sign-up period were accepted until September 30, 2009. An estimated 21,000 applications covering approximately 33 million acres were accepted. States receiving the most applications are Nebraska (2,683), Texas (1,879), and Missouri (1,821).[10] Total enrollment is capped at 12,769,000 acres annually.

One possible implementation issue with the new CSP could be similar to one faced by the old CSP, that of technical assistance cost restrictions. NRCS argued that because the 2002 farm bill placed a statutory 15% limit on technical assistance cost to administer CSP, program implementation was constrained. The new CSP could face similar implementation concerns owing to a 2008 farm bill requirement that the program achieve a national average program cost of $18 per acre. This per-acre cost includes all financial assistance paid to the producer, technical assistance, and any other expense associated with enrollment or participation in the program.

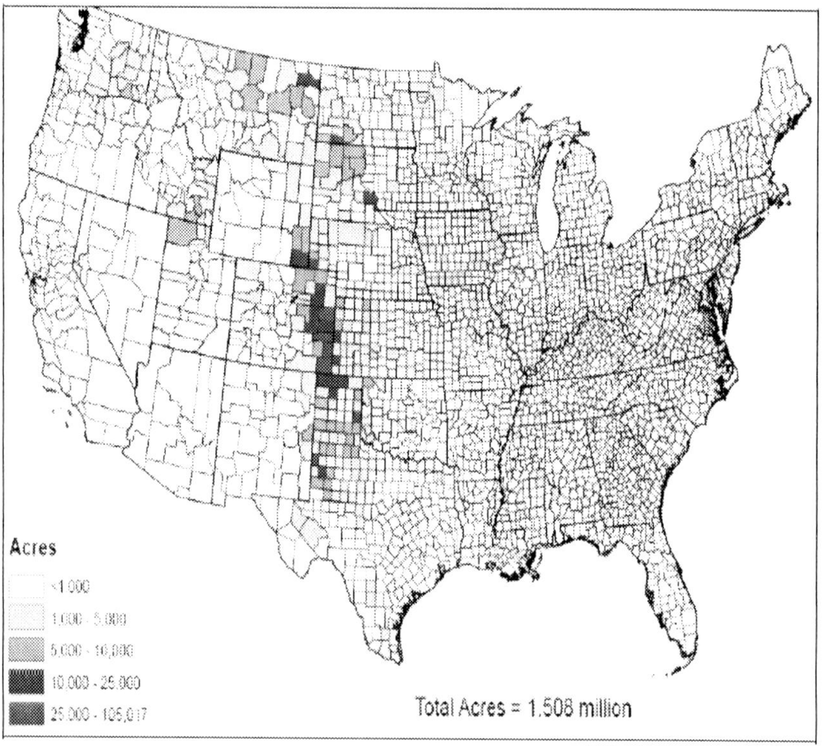

Source: USDA, Farm Service Agency

Figure 2. FY2009 Expiring CRP Acreage Eligible for Contract Extension

Table 1. EQIP Funded and Unfunded Applications and Funds Obligated

Fiscal Year	Total Applications Funded	Total Applications Unfunded	Percentage of Applications Funded	Funds Obligated (Financial Assistance, in thousands)
2000	16,249	37,712	30%	$139,606
2001	17,648	29,777	37%	$160,123
2002	19,817	70,495	22%	$322,193
2003	30,251	174,062	15%	$483,484
2004	46,413	135,394	26%	$718,150
2005	49,406	32,708	60%	$794,261
2006	41,190	32,633	56%	$787,968
2007	41,700	40,535	51%	$784,186
2008	48,116	23,803	67%	$924,221

Source: USDA, NRCS, *EQIP Contract and Funding Information*, http://www.nrcs.usda.gov/programs/eqip/.

Wildlife Habitat Incentives Program (WHIP)

The 2008 farm bill placed new limitations on lands eligible for enrollment in WHIP, which will likely reduce the number of eligible acres in traditionally high participation states. Language was added requiring WHIP to be used "for the development of wildlife habitat on private agricultural land, nonindustrial private forest land, and tribal lands." The addition of this language reverses the previous interpretation by NRCS, which extended eligibility to all privately owned land, tribal land, state/local government land, or federal land. By previously offering support for wildlife habitat projects on all land and aquatic areas, WHIP provided an assistance option to landowners who were unable to meet the specific eligibility requirements of other USDA conservation programs. The new requirement of agricultural production could shift financial assistance for WHIP contracts away from the traditionally large-allocation states in the Northeast to more agriculturally intensive states in the Midwest.

Wetlands Reserve Program (WRP)

The 2008 farm bill reauthorized WRP and increased the maximum enrollment to 3,041,200 acres, an increase of 766,200 acres above the previous cap established in the 2002 farm bill. The program receives mandatory funding through the Commodity Credit Corporation (CCC) and is estimated at $441 million for FY2009.[11] Between FY2003 and FY2007, WRP enrollment was reduced through annual appropriations bills. As Congress faces continued budget constraints, WRP could see similar reductions in the future.

One provision in the 2008 farm bill has generated considerable response from the environmental and conservation communities. Under Section 2203, Congress extended the length of time required to own the land before an applicant is eligible for the program from one year to seven years. The bill retained the exemptions to the ownership requirement, allowing a landowner to enroll land owned for less than seven years if the land was obtained (1) through inheritance; (2) through foreclosure; or (3) for any reason that gives NRCS assurance that the property was obtained for a purpose other than placing it in WRP. An investigation, conducted by the House Agriculture Committee, prior to the seven-year requirement found that NRCS "routinely ignored, or excluded its non-compliance with, the twelve-month ownership requirement" when in many cases partnering organizations would purchase the land for the explicit purpose of enrolling it into a WRP easement.[12] Many environmental organizations believe the seven-year requirement is excessive and will reduce the number of eligible lands and discourage enrollment. Supporters of the requirement point to the results of the House Agriculture Committee's investigation, which stated that NRCS "became a cash cow enabling partner organizations to acquire private lands at discount prices."[13]

Farmland Protection Program (FPP)

The 2008 farm bill (P.L. 110-246, Sec. 2401) reauthorized FPP and made significant changes. The enacted bill changes the program's purpose from protecting topsoil to protecting the land's agricultural use by limiting nonagricultural uses and including lands that promote state and local farmland protection. It also restructures the program to emphasize longer-term and renewable cooperative agreements, and changes the authority of the Secretary from *purchasing* conservation easements to *facilitating the purchase* of conservation easements. Some conservation organizations have praised the increased authority for partners, while others have expressed concerns that it could create inconsistencies within the program.

New Programs

In addition to the changes made to existing agricultural conservation programs, the enacted 2008 farm bill also expands the range of USDA conservation activities by creating several new programs, including a program expanding conservation activities in the Chesapeake Bay region, a new state grants program, a provision to limit production on native sod, and a provision promoting market-based approaches to conservation. The progress USDA has made in implementing these new programs varies and could be of interest for congressional oversight.

Chesapeake Bay Watershed Program

This program (Sec. 2605) is targeted at conserving and protecting the Chesapeake Bay and the water sources that make up the watershed. The bill authorizes $188 million in mandatory funding (FY2009-FY2012) to be used through existing conservation programs. Targeting funding to specific geographic regions was debated during the 2008 farm bill. The effectiveness of targeting federal conservation spending through identified states, regions, or watersheds is still a question when developing conservation policy. The Chesapeake Bay Watershed Program has received widespread support indicating that similar programs targeting other watersheds could be created by Congress in the future.

Voluntary Public Access and Habitat Incentives Program

Also referred to as the "Open Fields" program, this program authorizes state grants to encourage landowners to provide public access for wildlife-dependent recreation. The bill provides $50 million in mandatory funds (FY2009-FY2012) for the program. Despite authorized mandatory funding, USDA has not implemented this program as of the time of this report.

Sodsaver Provision

This 2008 farm bill provision (Sec. 12020) makes producers that plant an insurable crop (over 5 acres) on native sod ineligible for crop insurance and the noninsured crop disaster assistance (NAP) program for the first five years of planting. The conference agreement states that this provision may apply to virgin prairie converted to cropland in the Prairie Pothole National Priority Area (**Figure 3**), if elected by the state. USDA established a sign-up date of February 15, 2009, in which no governors opted to participate in the program. Additional opportunities to participate are possible though thought unlikely if the program remains voluntary.

Table 2. Conservation Programs Affected by Payment and Income Limitation Changes in the 2008 Farm Bill

Prior Law (P.L. 107-171, as amended)		Current Law (P.L. 110-246)	
Payment Limit	Income Limit	Payment Limit	Income Limit
AMA, CRP, and EQIP	AMA, Chesapeake Bay Watershed Program, CRP, CSP, EQIP, FPP, GRP, WRP, WHIP, and other programs under Title II of the 2002 farm bill	AMA, Chesapeake Bay Watershed Program, CRP, CSP, CCPI, EQIP, AWEP, FPP, GRP, WRP, and WHIP	AMA, AWEP, CRP, CSP, EQIP, FPP, GRP, GSWC, WRP, WHIP, and other programs under Title II of the2008 farm bill

Source: Income limits may be found under Section 1604 of the 2002 farm bill (P.L. 107-171) and Section 1604 the 2008 farm bill (P.L. 110-246). Payment limitations may be found under individual program sections under Title II of both enacted bills.

Notes: Listed programs include Environmental Quality Incentives Program (EQIP), Conservation Stewardship Program/Conservation Security Program (CSP), Wildlife Habitat Incentives Program (WHIP), Agricultural Water Enhancement Program (AWEP), Agricultural Management Assistance (AMA), Conservation Reserve Program (CRP), Wetlands Reserve Program (WRP), Farmland Protection Program (FPP), Grassland Reserve Program (GRP), Cooperative Conservation Partnership Initiative (CCPI), and Ground and Surface Water Conservation (GSWC).

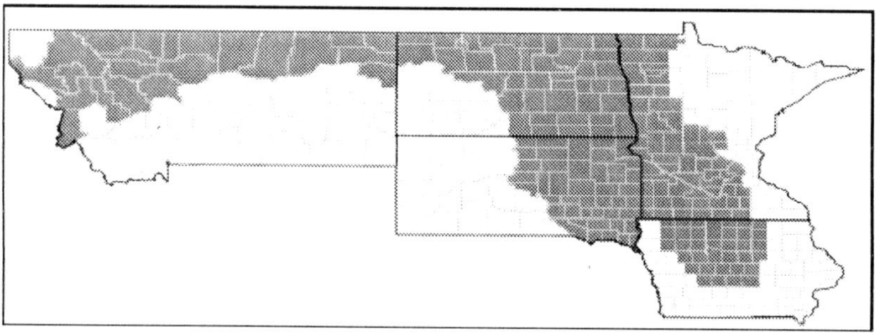

Source: USDA, RMA, Prairie Pothole National Priority Area, April 28, 2008, http://www.rma.usda.gov/data/ pothole/2008/all_states.pdf

Figure 3. Prairie Pothole National Priority Area

Environmental Services Markets

This new conservation provision (Sec. 2709) is intended to facilitate the participation of farmers and landowners in emerging environmental services markets, such as water and air quality, habitat protection, and carbon storage.[14] The inclusion of this provision could expand the scope of existing farmland conservation programs by facilitating the development of private-sector markets for a range of environmental goods and services from farmers and landowners (this is discussed further in the section on "Ecosystem Services Markets," below). Although the provision covers a range of farm and forestry services, it explicitly gives priority to first establishing guidelines related to participation in carbon markets.

The farm bill also directs USDA to establish a framework for developing consistent standards and processes for quantifying environmental services from the agriculture and forestry sectors, but does not authorize funding for this effort. In December 2008, USDA announced the creation of the Office of Ecosystem Services and Markets (OESM) and the creation of a federal government-wide Conservation and Land Management Environmental Services Board.[15] According to USDA , it intends to seek nominations for a federally chartered public advisory committee to advise the board.

Payment and Income Limitations

Two types of payment limits exist for conservation programs. One sets the maximum amount of conservation program payments that a person or legal entity can receive during a specified period of time The other sets the maximum amount of income that an individual can earn and still remain eligible for conservation program benefits (also know as the Adjusted Gross Income or AGI limit). Limitations on payments received through conservation programs were expanded in the 2008 farm bill. Prior to the 2008 farm bill, most conservation programs were affected by an income limitation, not a limitation on payments. Now, most programs are affected by both (see **Table 2**).

Payment Limits

Payment limits are the maximum amount of conservation program funding that a person or legal entity can receive during a specified period of time. As with commodity programs, payment limits for conservation programs are controversial because of issues relating to the size of operations receiving support and who should receive payments. The effect of payment limits varies by program and the conservation practices implemented.[16] Most conservation programs with higher payments tend to be skewed to farms and ranches with larger acreage because payments for many conservation practices are scaled by the number of acres on which that practice is applied.[17]

Supporters of payment limits are often advocates for smaller farms and opponents of large animal feeding operations. Most conservation programs provide a percentage of the cost to install conservation practices (known as cost-share) or implement site-specific management practices. As noted above, most of these payments are made on a per-acre applied basis, thereby skewing larger payments to contracts with more acres enrolled. Small farm advocates claim that this disproportionately benefits large agricultural producers by making less money available for small producers. Also, in the case of EQIP, cost-share assistance is provided for more expensive practices such as animal waste storage facilities in concentrated animal feeding operations (CAFOs).[18] Opponents of these animal operations cite the higher payment limit as "further subsidizing an environmentally destructive method of production."[19]

Those who oppose payment limits (or support higher limits) for conservation programs counter that conservation programs should focus on land with the greatest environmental need and not be limited to a price per participant. They argue that higher payment limits allow for greater environmental stewardship on farms and ranches, particularly larger operations with a greater land base, which may have greater natural resource concerns. Others claim that payment limits on restoration agreements could create a disincentive to enroll larger conservation easements, which are the most desirable. Because most conservation easement programs, namely WRP and GRP, enroll land that will also require restoration, a limit on restoration payments could reduce the enrollment of large acre tracts.

Income Limits

The AGI limit sets a maximum amount of income that an individual can earn and still remain eligible for program benefits. The 2008 farm bill made the AGI limitation for conservation programs higher than the AGI limitation for the commodity farm support programs; however, income limitations on conservation programs have been somewhat controversial. Previously, the AGI limit for both conservation and commodities programs was set at $2.5 million and had an exception if three-fourths of AGI was earned from farming sources. Now, if the three-year average of non-farm income AGI exceeds $1,000,000, no conservation program benefits are allowed. The exception to this limit is if two-thirds of the three-year AGI was earned from farming sources. In addition, this limitation may be waived by USDA on a case-by-case basis for the protection of environmentally sensitive land of special significance. In general, the AGI limit for conservation programs is higher than that for commodity programs to encourage environmental stewardship on farms and ranches, particularly larger operations that may have greater natural resource problems.

Supporters of AGI limits believe that tighter limits benefit small producers and gain additional public support for all agricultural programs through fiscal responsibility. Opponents of AGI limits on conservation programs believe that if there are greater conservation benefits provided to the general public, then, irrespective of wealth, a producer's enrollment is good for the general public.

Conservation Funding and Oversight

The majority of conservation program funding is mandatory and funded through the Commodity Credit Corporation (CCC).[20] Other conservation programs, mostly technical assistance, are discretionary and funded through annual appropriations. Reductions in mandatory funding and an increase in earmarks within discretionary funding, as well as fund accountability and quantifying environmental benefits of conservation practices, could be issues for Congress.

Spending Limits on Mandatory Funding

The 2002 farm bill authorized significant increases in mandatory funding for many conservation programs. Unlike the discretionary conservation programs, which must be funded through the annual appropriations process, mandatory programs have an authorized level of funding (or acreage enrollment) that is available unless reduced to smaller amounts in the appropriations process. If appropriators do not set a spending limit or reduce the authorized level, then the program receives the authorized level of funding. Mandatory conservation programs are usually authorized through omnibus farm bills for four to five years. Mandatory funds are provided by the CCC, USDA's financing entity for many other agricultural programs and export subsidies.

Despite the increase in mandatory funding authority, many conservation programs have been reduced or capped through annual appropriations acts since FY2003. Many of these spending reductions were at the request of the Bush Administration. The mix of programs and amounts of reduction have varied from year to year. Some programs, such as the CRP, have not been reduced by appropriators in recent years, while others, such as EQIP, have been repeatedly reduced below authorized levels. EQIP's authorized mandatory funding was reduced by almost $1.16 billion from FY2005 through

FY2010. Even with these reductions, total mandatory funding for conservation programs has remained relatively constant at a total over $4 billion annually.

Although some titles in the 2008 farm bill received a reduction in funding over the next five to ten years, the conservation title received increased funding. Many supporters of conservation programs viewed this as a victory during the farm bill debate and were surprised to see that the recent FY2010 President's budget requested a reduction of over $500 million in mandatory funding from many conservation programs.[21] Advocates for these programs contend that these limitations are significant changes from the intent of the farm bill, which they say compromise the programs' ability to provide the anticipated magnitude of benefits to producers and the environment. Others, including those interested in reducing agricultural expenditures or in spending the funds for other agriculture purposes, counter that, even with these reductions, overall funding for conservation has not been reduced. With the 111th Congress facing tighter budget constraints, it is possible for conservation programs to face spending reductions either in the appropriations process or if budget reconciliation is required.[22]

Congressionally Directed Projects

The Watershed and Flood Prevention Operations program has received discretionary funding for an increasing number of congressionally directed projects (commonly referred to as earmarks). The program is not authorized at a specified amount. Annual appropriations reached a high of $200 million, including a $94 million supplemental, in FY2002 and have declined steadily to $24.3 million in FY2009.[23] The program also received $145 million as part of the American Recovery and Reinvestment Act of 2009 (ARRA, P.L. 111-5). As appropriated funding decreased, the number of earmarks increased, reaching 97% of appropriated funding in FY2009 and 74% in FY2010. The past five presidential budget requests, including the FY2010 request, have sought zero funding for the program, citing the number of earmarks. Currently there are approximately 300 unfunded authorized watershed projects totaling over $1 billion.

Technical Assistance

Two issues associated with technical assistance have been whether NRCS has the capacity to meet the growing technical assistance demand and how technical assistance costs should be funded for mandatory programs.

The 2002 farm bill (P.L. 107-171, Sec. 2701) dramatically increased financial assistance in conservation programs. This increase in financial assistance also led to an increased demand for technical assistance.[24] The 2002 farm bill allowed NRCS to augment its technical assistance capacity by allowing producers to use approved third parties to provide assistance. As of August 2008, approximately 1,700 individuals and businesses were certified and registered as third-party providers with NRCS.[25] Despite this increase in capacity, NRCS, like other organizations, is facing personnel attrition in "mission critical occupations" (most of which provide technical assistance).[26] The 2008 farm bill further increased funding for mandatory conservation programs, which could cause the demand for technical assistance to increase, placing additional strain on the current capacity.

Congress has addressed the issue of how to fund technical assistance in mandatory programs in the past and will likely do so again in the future. The central issue is whether NRCS has the authority to pay for technical assistance for mandatory programs using the mandatory programs' funding.[27] Previous conflicting interpretations by the Office of Management and Budget resulted in Congress providing clarifying language through enacted legislation in 2002, 2005, and 2008.[28] The clarifying language appeared to resolve the issue until recently. The 2008 farm bill reauthorized and amended other mandatory conservation programs not originally authorized under the 1985 farm bill.[29] Because previous language only addressed those programs authorized under the 1985 farm bill, the issue of where technical assistance funding should originate for non-1985 farm bill programs was once again raised. On February 17, 2009, Congress enacted clarifying language for this issue in Section 103 of the American Recovery and Reinvestment Act of 2009 (P.L. 111-5).[30] The new language allows all mandatory programs established or amended in the 2008 farm bill, excluding those in Title I, to use CCC funds for administrative expenses and technical assistance. Congress will likely need to address this issue again because the new language only covers FY2009 and FY2010, while funding authority for most farm bill programs expires at the end of FY2012.

Emergency Conservation Programs

Emergency programs are implemented most years when funding is provided in response to natural disasters. The most recent funding was provided in response to flooding and drought conditions in the spring and summer of 2008. Congress provided $390.5 million for the Emergency Watershed Program (EWP) and $89.4 million for the Emergency Conservation Program (ECP) in Title III of the 2008 Supplemental Appropriations Act (P.L. 110-252). Both programs received additional funding through the Consolidated Security, Disaster Assistance, and Continuing Appropriations

Act of 2009 (P.L. 110-329), with the EWP receiving $100 million and the ECP receiving $115 million. Recent flooding, wildfires, and drought conditions have increased demand beyond available funding for both programs. History indicates that funding could be sought for both programs before the end of the fiscal year through an emergency supplemental, pending a natural disaster.

Financial Audit and Contract Administration

Most USDA agencies are covered by a consolidated financial statement audit for the entire department. In FY2007, the Office of Management and Budget (OMB) required NRCS to conduct a stand-alone audit for the FY2008 financial statement. In FY2008, NRCS contracted with a private auditing firm, KPMG, to review the agency's financial statements in its first standalone audit. KPMG was unable to provide an opinion on the agency's financial statements because NRCS could not document or support its transactions or account balances.[31]

The FY2008 audit was NRCS's first attempt at a comprehensive financial statement audit. The lack of an opinion from the auditing firm was a result of several problems—namely, NRCS was unable to provide material in support of transactions and account balances with respect to obligations, accrued expenses and undeliverable orders, and unfilled customer orders. According to NRCS, the agency has taken corrective actions to address the deficiencies found in the audit, including training over 300 employees in accounting principles, developing an automated tool for general ledger accounts, and performing quality assurance reviews on over 160,000 open obligations.[32] An audit of the FY2009 financial statement is currently in progress.

The 111th Congress has already taken action on this issue. On March 25, 2009, the House Committee on Agriculture's Subcommittee on Conservation, Credit, Energy, and Research held a hearing to review the administration of conservation contracts. As additional audit reports are issued, additional hearings are possible. Also, in the FY2009 Omnibus Appropriations Act, Congress provided $10 million in Conservation Technical Assistance (CTA) funds to "strengthen the agency's program and financial management capabilities."[33] According to the committee print, NRCS is expected to use the funds to enhance the agency's budgeting, accounting, contracting, and information technology systems. The agency is expected to issue a report to Congress detailing the use of these funds. The committee print also requires NRCS to issue a report to Congress regarding the 2008 audit and any corrective actions being taken.

Conservation Effects Assessment Project (CEAP)

Following the significant increase in funding for conservation programs in the 2002 farm bill, USDA initiated a project to measure the environmental benefits of many of these programs. The project is a multi-agency effort known as the Conservation Effects Assessment Project (CEAP). CEAP's stated purpose is to aid policymakers in developing new conservation programs and help existing conservation program managers implement programs more effectively and efficiently to meet the goals of Congress and the Administration.[34]

Now in its sixth year at USDA, CEAP continues to be supported by many in the conservation community as a much-needed evaluation system to ensure that conservation programs produce the desired environmental and production outcomes.[35] Although the project is not legislatively mandated, many believe its potential outcomes could prove useful in shaping future policy debates surrounding environmental issues in the 111th Congress, namely environ mental services markets.

Despite some of the strides made to quantify the environmental benefits of conservation practices through CEAP, there are possible limitations to its use, namely the availability of data. To date, the depth of information gathered through this project is enormous, yet very little has been made available to the public or to other federal agencies. Despite some frustration generated by reporting delays, some in the scientific community support what they say is a meticulous in-depth assessment and evaluation system, claiming it is necessary for statistical credibility.[36] As of this report, the first initial cropland regional assessment has not been released.

Emerging and Continuing Issues

Emerging and continuing issues for agricultural conservation in the 111th Congress revolve around ecosystem services markets, climate change, and bioenergy. The evolving debate over climate change legislation has provoked several questions regarding the role of agriculture and forestry. Other environmental concerns for agriculture, such as concentrated animal feeding operation (CAFO) regulations, greenhouse gas emission reporting for livestock producers, and wetlands mitigation, could lead to expanded opportunities for many conservation efforts. Production pressures generated by corn-based ethanol have also had an ongoing impact on certain conservation programs.

Ecosystem Services Markets

The participation of agriculture and forestry in emerging ecosystem services markets is gaining wide support within the farm community and its supporting organizations and agencies, as well as among the regulatory agencies and some environmental groups.[37] Market-based approaches are often viewed as encompassing broader societal benefits by complementing existing agricultural conservation programs and evolving regulatory approaches intended to address environmental improvements in the farm and forestry sectors. Among the principal questions regarding the inclusion of ecosystem services provisions as part of any major legislative initiative is whether the agriculture and forestry sectors can effectively provide environmental goods and services along with the more traditional food, fiber, and other services these sectors already provide.

Ecosystem goods and services are the benefits society obtains from the environment and ecosystems, both natural and managed, such as water filtration, flood control, provision of habitat, carbon storage, and many others.[38] In most cases, these constitute "free services," since landowners and managers are not compensated in the marketplace. The continued degradation of these services over time has generated interest in creating markets, developed either through regulation or voluntarily, so that providers of ecosystem services can be compensated in private markets for the services they provide. This could offer a potential business opportunity to the agriculture and forestry sectors, which may be able to provide for such services and participate in the market, for example, by creating, restoring, and preserving function and value in a natural resources area, or by capturing and storing carbon before gases that contribute to global climate change are released into the atmosphere. These services would be in addition to the food and fiber services traditionally supplied by the agriculture and forestry sectors.[39]

Conservation practices on agricultural land provide one of the main vehicles through which producers may develop environmental credits and benefits to be sold or traded in a market-based system. Many existing USDA conservation programs offer incentives through financial and technical assistance to implement these practices. For example, cropland tillage practices such as reduced/medium-till, no-till, and ridge/strip-till practices sequester carbon that could be sold through a carbon offset program. Conservation programs such as EQIP, AMA, CSP, and CTA provide incentives for farmers to install these practices. Programs such as WRP offer incentives for wetland restoration, the benefits of which could then be used in a wetlands mitigation program. Other conservation practices such as riparian buffers, setbacks, wind breaks, and buffer strips offered under EQIP, CRP, CSP and WHIP create water quality improvements that could be traded in local water quality credit programs.

The inclusion of these market provisions in legislation has raised procedural and implementation questions as Congress continues to consider the role of the agriculture and forestry sectors in environmental legislation.[40] One key question is how USDA will respond to several existing barriers that may prevent the development of ecosystem goods and services markets involving the farm and forestry sectors, such as participation challenges, and issues of measuring and valuing credits, monitoring, and enforcement, among others.[41] It is also unclear how USDA will bridge existing conservation program benefits with ecosystem services markets, and what direct role the new USDA Office of Ecosystem Service Markets (see "Environmental Services Markets," above) plans to play.

Climate Change

Among the many questions for Congress regarding the topic of climate change is what role the agriculture and forestry sectors will play; and, if they are allowed to participate, how concerns regarding their participation will be resolved. Farm and forestry activities both emit and sequester greenhouse gases (GHG) from the atmosphere.[42] As reported by the Environmental Protection Agency (EPA), the agriculture and forestry sectors currently account for 6%-8% of estimated total U.S. GHG emissions annually.[43] However, combined carbon sequestration on farm and forested lands is currently estimated to mitigate about 11% of total annual GHG emissions in the United States. The potential to reduce emissions and sequester carbon on agricultural lands is reportedly much greater than current rates.

Congress is considering a range of climate change policy options, including greenhouse gas (GHG) emission reduction programs that would either mandate or authorize a cap-and-trade program to reduce GHG emissions.[44] In general, the current legislative proposals would not require emission reductions in the agriculture and forestry sectors. The drafters of some proposals are reluctant to include opportunities for agriculture and forestry participation due to uncertainties regarding measuring, monitoring, and permanence of agriculture and forestry activities.[45] Others in Congress would like a GHG program to include provisions allowing farm and forestry landowners to receive emissions allowances (or credits) to generate carbon offsets.

Most land management and farm conservation practices can help reduce GHG emissions and/or sequester carbon, including land retirement, conservation tillage, soil management, and manure and animal feed management, among other practices. Many of these practices are encouraged under existing USDA conservation programs that provide financial and technical assistance to farmers, such as EQIP, CRP, CSP, CTA, and WHIP, among others. However, uncertainties are associated with implementing these types of practices depending on site-specific conditions, the type of practice, how well it is implemented, the length of time a practice is undertaken, and available funding, among other factors.

USDA has already expanded some of its existing conservation programs to further encourage emission reductions and carbon sequestration. For example, many of the practices encouraged under EQIP and CSP reduce net emissions. Programs such as CTA, AMA, EQIP, and CSP list a reduction in emissions as a national priority for the program, which affects the funding and ranking of projects. Under CRP, USDA has modified how it scores and ranks offers to enroll land in CRP in order to place greater weight on installing vegetative covers that sequester carbon. USDA also has an initiative under CRP's continuous enrollment provision to plant up to 500,000 acres of bottomland hardwoods, which are among the most productive U.S. lands for sequestering carbon.[46] Recently, USDA has recognized the potential credits generated by these conservation programs and has removed any claim on the credits through recent changes to many of the program rules.[47] Whether or not benefits generated from existing conservation programs would be considered eligible carbon offsets under current legislative proposals varies.

Congress has already taken steps to address some of the challenges associated with measuring carbon changes from forested and agricultural lands and practices. The 2008 farm bill includes provisions that could expand the scope of existing land-based conservation and other farm bill programs by providing incentives to encourage farmers and landowners to sequester carbon and reduce emissions associated with climate change and participate in markets for carbon storage (see "Environmental Services Markets" section, above). The Energy Independence and Security Act of 2007 (P.L. 110-140, Sec. 712) directs the Secretary of the Interior to develop a methodology to assess carbon sequestration and emissions from ecosystems. This methodology is to cover measuring, monitoring, and quantifying GHG emissions and reductions, and provide estimates of sequestration capacity and the mitigation potential of different ecosystem management practices. Congress continues to face the question of whether its current authorized activities (and proposed activities) provide adequate and sufficient guidelines for accurately measuring carbon levels from forest and agricultural activities.

Energy Effects

Renewable energy and energy conservation continue to be topics of concern as Congress seeks to reduce the national use of fossil fuels. Renewable energy is energy generated through natural means, including wind, solar, geothermal, hydroelectric, and biomass, among others. With each renewable energy source comes benefits and consequences. Energy sources closely tied to the agriculture and forestry sectors are biofuels derived from corn-based ethanol and cellulosic feedstock. Federal support for ethanol through mandated levels of consumption, grants and loan guarantees, tax credits, tariffs on imports, and research and development projects has generated some controversy.[48] Critics say that important and unwanted environmental consequences resulting from the production of biofuels should be considered when evaluating renewable energy options.

A continued expansion of corn-based ethanol production could have significant consequences for traditional U.S. agricultural crop production and rural economies. Supporters of an expanded renewable fuels standard (RFS, the statutorily required use of renewable fuels) claim that increased biofuels production and use would produce enormous agricultural and rural economic benefits by increasing farm and rural incomes and generating substantial rural employment opportunities.[49] However, large-scale shifts in agricultural production activities will likely also have important regional economic consequences that have yet to be fully considered or understood. As corn prices rise, so too does the incentive to expand corn production to more marginal soil environments. Further, corn production is among the most energy-intensive of the major field crops. This shift could have important and unwanted environmental consequences due to possible increases in fertilizer and chemical use and soil erosion.

Ethanol and biodiesel produced from cellulosic feedstocks, such as prairie grasses and fast-growing trees, have the potential to improve the energy and environmental effects of U.S. biofuels. A key potential benefit of cellulosic feedstocks is that they can be grown without the need for chemicals. Reducing or eliminating the need for chemical fertilizers could address one of the largest energy inputs for corn-based ethanol production. However, additional concerns about cellulosic feedstocks exist, including concerns that required increases in per-acre yields to obtain economic feasibility could require the use of fertilizers; that availability of sufficient feedstock supply is limited and expansion could generate additional land use pressures for expanded production; and that development of harvesting machinery and technology lags behind that of traditional corn-based ethanol production. In addition to these concerns, some groups say that other potential environmental drawbacks associated with cellulosic fuels should be addressed, such as the potential for soil erosion, increased runoff, the spread of invasive species, and disruption of wildlife habitat.[50]

Although the 2008 farm bill expands many existing agricultural conservation programs to encourage producers to adopt energy efficiency measures and produce renewable energy feedstocks, it does not address the potential environmental consequences that could result from biofuels production. Expanding existing conservation efforts or creating new ones can serve as a potential countermeasure to the increased environmental pressures generated by biofuel production. The extent to which this is or could be done is not clear.

The EPA has proposed changes to the RFS program, as required by the Energy Independence and Security Act of 2007 (EISA, P.L. 110-140).[51] Two areas of this proposed rulemaking have caused concerns among those in the U.S. agriculture sector: the EISA biomass definition and the requirement that EPA consider so-called "indirect land use" effects when calculating GHG emissions associated with advanced biofuels. Recent amendments to the Waxman/Markey climate change and energy legislation (H.F.. 2454) would change the EISA biomass definition to the definition in the 2008 farm bill (P.L. 110-246) and prevent EPA (for at least six years) from including emissions from international indirect land use changes when weighing biofuels' carbon footprint under the 2007 RFS. On June 29, 2009, the House passed H.R. 2454, which also included a new program for agriculture- and forestry-related offsets within USDA.[52] Other energy and offset program proposals are currently being considered in the Senate.[53]

Appendix. USDA Conservation Programs: Funding

Table A-1. USDA Agricultural Conservation Program Funding Levels, FY2009 and FY2010 (listed by FY2009 funding level, in millions)

Program	Mandatory/ Discretionary	FY2009 Available	FY2010 Available
Conservation Reserve Program (CRP)	Mandatory	$1,888 est. (39.2 million acres)	$1,872 est. (32 million acres)
Environmental Quality Incentives Program (EQIP)	Mandatory	$1,067 ($1,337 authorized)	$1,180 ($1,450 authorized)
Conservation Operations (includes Conservation Technical Assistance, Survey, Soil Survey, Grazing Lands Conservation Initiative, Plant Materials Centers, Conservation Compliance, Sodbuster, and Swampbuster)	Discretionary	$730	$888
Conservation Stewardship Program (CSP)	Mandatory	$569 est. (12.5 million acres annually)	$752 est. (12.5 million acres annually)
Wetlands Reserve Program (WRP)	Mandatory	$417 est. (3.04 million acres)	$671 est. (3.04 million acres)
Emergency Watershed Program (EWP)	Discretionary	$245	$0
Watershed and Flood Prevention Operations	Discretionary	$169	$30
Farmland Protection Program (FPP)	Mandatory	$121	$150
Emergency Conservation Program (ECP)	Discretionary	$115	$0
Watershed Rehabilitation	Mandatory and Discretionary	$90	$41.1 discretionary, $0 mandatory ($165 authorized)

Table A-1. (Continued)

Program	Mandatory/ Discretionary	FY2009 Available	FY2010 Available
Wildlife Habitat Incentives Program (WHIP)	Mandatory	$85	$85
Agricultural Water Enhancement Program (AWEP)	Mandatory	$73	$73
Grassland Reserve Program (GRP)	Mandatory	$63	$78
Resource Conservation and Development (RC&D)	Discretionary	$51	$51
Chesapeake Bay Watershed Program	Mandatory	$23	$43
Healthy Forests Reserve Program (HFRP)	Mandatory	$9.8	$9.8
Agricultural Management Assistance (AMA)	Mandatory	$7.5 (conservation only)	$7.5 (conservation only)
Voluntary Public Access and Habitat Incentive Program	Mandatory	$50 (total for FY2009-2012)	$50 (total for FY2009-2012)

Source: Division A, Title II of the Omnibus Appropriations Act, 2009 (P.L. 111-8); Division A, Title I of the American Recovery and Reinvestment Act of 2009 (P.L. 111-5); Title I and II of the Food, Conservation, and Energy Act of 2008 (P.L. 110-246); and Division B, Title I, Chapter I of the Consolidated Security, Disaster Assistance, and Continuing Appropriations Act, 2009 (P.L. 110-329). Funding estimates for acreage based programs are taken from CBO Baseline projections, March 2009; and USDA, CCC, *Commodity Credit Corporations - Commodity Estimates Book*, FY 2010 Mid-Session Review, August 25, 2009, http://www.fsa.usda.gov/Internet/FSA_File/msr10_commodity.

Notes: FY2009 available column represents the actual appropriated or authorized level of funding. The total includes funding from the American Recovery and Reinvestment Act of 2009 (P.L. 111-5), which provided additional funding for Watershed and Flood Prevention Operations, Watershed Rehabilitation, and Emergency Watershed Program (in the form of floodplain easements). Both EWP and ECP received funding in the Consolidated Security, Disaster Assistance, and Continuing Appropriations Act, 2009 (P.L. 110-329). FY2010 authorized limits are provided in title I and II of the Food, Conservation, and Energy Act of 2008 (P.L. 110-246).

End Notes

[1] Conservation and Trade Act of 1990 (P.L. 101-624), Federal Agricultural Improvement and Reform Act of 1996 (P.L. 104-127), Farm Security and Rural Investment Act of 2002 (P.L. 107-171), and Food, Conservation, and Energy Act of 2008 (P.L. 110-246).

[2] The CCC is the funding mechanism for the mandatory payments that are administered by various agencies of USDA.

[3] For more information on ARS projects, see http://www.ars.usda.gov/Research/Research.htm; for ERS projects, see http://www.ers.usda.gov/ Browse/view.aspx? subject=Natural ResourcesEnvironment; and for FS projects, see http://www.fs.fed.us/projects/.

[4] For additional information on agricultural conservation programs, see CRS Report R40763, *Agricultural Conservation: A Guide to Programs*.

[5] For additional information on conservation provisions in the farm bill, see CRS Report RL34557, *Conservation Provisions of the 2008 Farm Bill*.

[6] For further information on CRP, see CRS Report RS21613, *Conservation Reserve Program: Status and Current Issues*.

[7] Annual appropriations reduce funding for several other agriculture mandatory programs as a means of meeting overall budget targets. EQIP, however, has taken the largest reductions of all conservation programs.

[8] Many of CSP's implementation issues centered around additional eligibility and enrollment criteria added by USDA but not specified in the 2002 farm bill. These issues included implementing CSP in certain watersheds on a rotating basis; requiring producers to meet strict eligibility criteria; sorting producers' applications into enrollment categories; and reducing certain CSP payments. USDA justified the restrictive actions because congressional authorizers crafted an entitlement program with unlimited funding, but congressional appropriators limited that funding. For additional information on CSP, see CRS Report RS21740, *Conservation Security Program: Implementation and Termination*.

[9] USDA, CCC, "Conservation Stewardship Program," 74 *Federal Register* 144, July 29, 2009. An amendment was issued in September extending the public comment period. USDA, CCC, "Conservation Stewardship Program," 74 *Federal Register* 181, September 21, 2009.

[10] NRCS, *Conservation Stewardship Program Sign-up Application Report*, October 9, 2009, http://www.nrcs.usda.gov/programs/new_csp/special_pdfs/CSP_Sign-Up_Application_Report_10-09-09.pdf.

[11] U.S. Congressional Budget Office, CBO January 2009 Baseline for CCC, FCIC, and Conservation, January 2009.

[12] U.S. Congress, House Committee on Agriculture, Subcommittee on Conservation, Credit, Energy, and Research, *Testimony from John Jurich, Investigator, Committee on Agriculture, U.S. House of Representatives*, subcommittee hearing to review the USDA administration of conservation program contracts, 111th Cong., 1st sess., March 25, 2009, http://agriculture.house.gov/testimony/111/h032509/Jurich.pdf.

[13] Ibid.

[14] For more information on the Environmental Services Markets provision in the 2008 farm bill, see CRS Report RL34042, *Provisions Supporting Ecosystem Services Markets in U.S. Farm Bill Legislation*.

[15] USDA, "USDA Announces New Office of Ecosystem Services and Markets," press release, December 18, 2008, *http://www.usda.gov/wps/* portal?contentidonly =true & contentid= 2008/12/0307.xml.

[16] Payment limits vary from $300,000 for any six-year period for EQIP to $50,000 annually for WHIP payments.

[17] Soil and Water Conservation Society and Environmental Defense, *Environmental Quality Incentives Program (EQIP)—Program Assessment*, March 2007, *http://www.swcs.org/* documents/ filelibrary/EQIP_assessment.pdf.

[18] The 2002 farm bill increased the payment limit from its 1996 farm bill levels of $10,000 for any fiscal year and $50,000 for any multi-year contract, to a limitation on total payments to $450,000. The 2002 farm bill also required that 60% of EQIP funds should be made available to payments for practices relating to livestock production.

[19] Elanor Starmer, "Industrial Livestock at the Taxpayer Trough: How Large Hog and Dairy Operations are Subsidized by the Environmental Quality Incentives Program," Campaign for Family Farms and the Environment, December 2008, p. 3.

[20] The CCC is the funding mechanism for the mandatory payments that are administered by various agencies of USDA.

[21] The FY2010 President's budget requested reductions in the following conservation programs receiving mandatory funding: AMA ($5 million reduction), EQIP ($250 million reduction), FPP ($30 million reduction), HFRP ($5 million reduction), WHIP ($43 million reduction), and WRP (139,000 acres reduction).

[22] For more information on the reconciliation process in general, see CRS Report RL30458, *The Budget Reconciliation Process: Timing of Legislative Action.*

[23] Much of the decline had been at the request of the USDA. Source: Agriculture, Rural Development, Food and Drug Administration, and Related Agencies Appropriations Act, 2002 (P.L. 107-76); 2002 Supplemental Appropriations Act for Further Recovery From and Response to Terrorist Attacks on the United States (P.L. 107-206); and Omnibus Appropriations Act, 2009 (P.L. 111-8).

[24] Most conservation programs that include financial assistance require a certain amount of technical assistance. According to Section 2706 of the 2008 farm bill (P.L. 110-246), the purpose of technical assistance within farm bill programs is to "provide eligible participants with consistent, science-based, site-specific practices designed to achieve conservation objectives on land active in agriculture, forestry, or related uses."

[25] NRCS reports expending $217 million for technical assistance provided through a third-party between October 2003 through September 2008. See USDA, NRCS, "Technical Service Provider Assistance," 74 *Federal Register* 2800, January 16, 2009.

[26] USDA, NRCS, "Human Capital Strategic Plan, 2006-2010," at *http://www.nrcs.* usda.gov/about/humancapital/.

[27] Unlike most government agencies, NRCS does not receive a salaries and expenses appropriation. Most of the funding for employees and overhead (including what is known as technical assistance) is appropriated through the Conservation Operations account, which is discretionary. OMB originally determined that technical assistance for mandatory programs would be funded through annual discretionary appropriations and funding from larger mandatory programs. Congress disagreed and passed legislation stating that funding for a mandatory program's technical assistance could only come from that mandatory program's funding authority.

[28] Section 2701 of the 2002 farm bill (P.L. 107-171) provided that technical assistance in support of each mandatory program authorized under the 1985 farm bill, as amended, come from the funding provided by the CCC for that program P.L. 108-498, further clarified the funding source for technical assistance in mandatory conservation programs and specified that funding for technical assistance cannot be transferred among the mandatory funded programs, starting in FY2005. The language was reauthorized in the 2008 farm bill (P.L. 110-246).

[29] For example, the Healthy Forest Reserve Program authorized in section 502 of the Healthy Forest Restoration Act of 2003 (16 U.S.C. 6572), the Agricultural Management Assistance Program authorized in section 524 of the Federal Crop Insurance Act (7 U.S.C. 1524), and the Chesapeake Bay Watershed Program authorized in section 2605 of the Food, Conservation, and Energy Act of 2008 (16 U.S.C. 3839bb-3).

[30] For more information on the ARRA see, CRS Report R40160, *Agriculture, Nutrition, and Rural Provisions in the American Recovery and Reinvestment Act (ARRA) of 2009.*

[31] For a copy of the FY2008 NRCS audit report by USDA's OIG, see http://www.usda.gov/oig/webdocs/10401-2-FM.pdf.

[32] U.S. Congress, House Committee on Agriculture, Subcommittee on Conservation, Credit, Energy, and Research, *Testimony of Dave White, Chief, Natural Resources Conservation Service, USDA*, Hearing to review the USDA administration of conservation program contracts, 111th Cong., 1st sess., March 25, 2009.

[33] U.S. Congress, House Committee on Appropriations, Omnibus Appropriations Act, 2009, Division A—Agriculture, Rural Development, Food and Drug Administration, and Related Agencies Appropriations Act 2009, committee print, 111th Cong., 1st sess., March 2009 (Washington: GPO, 2009).

[34] Soil and Water Conservation Society, Final Report from the Blue Ribbon Panel Conducting an External Review of the U.S. Department of Agriculture Conservation Effects Assessment Project, Ankeny, IA, 2006, http://www.swcs.org/ documents/filelibrary/ advocacy_publications/CEAP_Final_Report.pdf.

[35] Lisa F. Duriancik, Dale Bucks, and James P. Dobrowolski et al., "The First Five Years of the Conservation Effects Assessment Project," *Journal of Soil and Water Conservation*, vol. 63, no. 6 (Nov/Dec 2008), p. 185A.

[36] "Quantifying Practices with the Conservation Effects Assessment Project (CEAP)," panel discussion at the 2009 USDA Agriculture Outlook Forum, Arlington, VA, February 26, 2009.

[37] In March 2009, Secretary of Agriculture Thomas Vilsack indicated that traditional agricultural support payments will continue to be met with criticism and many producers should seek additional income from market-based ecosystem services. U.S. Congress, House Committee on Appropriations, Subcommittee on Agriculture, Rural Development, Food and Drug Administration, and Related Agencies, *Statement by Thomas Vilsack, Secretary of Agriculture*, Submitted Remarks, 111th Cong., 1st sess., March 31, 2009. http://www.usda.gov/wps/portal?contentidonly=true& contentid=2009/03/0077.xml

[38] These may also be referred to as ecosystems services. See, for example, World Resources Institute, *Millennium Ecosystem Assessment, Ecosystems and Human Well-being*, 2005, http://www.millenniumassessment.org/en/ index.aspx.

[39] For additional information, see CRS Report RL34042, *Provisions Supporting Ecosystem Services Markets in U.S. Farm Bill Legislation*.

[40] For more information on the role of agriculture in climate change policy, see CRS Report RL33898, *Climate Change: The Role of the U.S. Agriculture Sector*. For other general information on climate change issues, see CRS Report RL34513, *Climate Change: Current Issues and Policy Tools*.

[41] For more information, see USDA, *2007 Farm Bill, Conservation and Environment Theme Paper*, June 2006, at http://www.usda.gov/documents/FarmBill07consenv.pdf; and M. Ribaudo and C. Jones, "Environmental Credit Trading: Can Farming Benefit," *Amber Waves*, USDA's Economic Research Service, Feb. 2006.

[42] GHG emissions from agriculture are associated with livestock operations (as part of the natural digestive process of animals and manure management) and crop production (soil management, commercial fertilizer and manure application, and production of nitrogen-fixing crops). The two key types of GHG emissions are methane (CH_4) and nitrous oxide (N_2O). Estimated emissions are expressed on a CO_2-equivalent basis. See CRS Report RL33898, *Climate Change: The Role of the U.S. Agriculture Sector*.

[43] EPA, *Inventory of U.S. Greenhouse Gas Emissions and Sinks: 1990-2007*, April 2009, http://epa.gov/climatechange/ emissions/usinventoryreport.html.

[44] For more information on the current GHG emissions legislation, see CRS Report R40556, *Market-Based Greenhouse Gas Control: Selected Proposals in the 111th Congress*; and CRS Report RL34436, *The Role of Offsets in a Greenhouse Gas Emissions Cap-and-Trade Program: Potential Benefits and Concerns*.

[45] For more information on issues associated with measuring and monitoring carbon, see CRS Report RS22964, *Measuring and Monitoring Carbon in the Agricultural and Forestry Sectors*.

[46] For more information on these programs, see CRS Report R40763, *Agricultural Conservation: A Guide to Programs*.

[47] The following program rules include a section recognizing the credits generated by programs and asserting no direct or indirect claim on these credits: EQIP (§1466.36, 74 *Federal Register* 2317), WRP (§1467.20, 74 *Federal Register* 2336), AMA (§1465.36, 73 *Federal Register* 70256), GRP (§1415.10, 74 *Federal Register* 3875), FPP (§1491.21, 74 *Federal Register* 2822), WHIP (§636.21, 74 *Federal Register* 2800), CRP (§1410.63(6), 68 *Federal Register* 24845), and HFRP (§625.8, §74 *Federal Register* 1967).

[48] For more information on ethanol policy issues, see CRS Report R40488, *Ethanol: Economic and Policy Issues*.

[49] For more information on the renewable fuels standard, see CRS Report R40155, *Selected Issues Related to an Expansion of the Renewable Fuel Standard (RFS)*.

[50] For more information on cellulosic biofuels, see CRS Report RL34738, *Cellulosic Biofuels: Analysis of Policy Issues for Congress*.

[51] 74 *Federal Register* 99, 24904-25143, May 26, 2009. EISA significantly expanded the RFS established in the Energy Policy Act of 2005 (P.L. 109-58). The RFS now requires the use of 9.0 billion gallons of renewable fuel in 2008, increasing to 36 billion gallons in 2022.

[52] For more information on H.R. 2454, see CRS Report R40643, *Greenhouse Gas Legislation: Summary and Analysis of H.R. 2454 as Passed by the House of Representatives*.

[53] For more information on these requirements and related RFS issues, see CRS Report R40155, *Selected Issues Related to an Expansion of the Renewable Fuel Standard (RFS)*; and CRS Report R40460, *Calculation of Lifecycle Greenhouse Gas Emissions for the Renewable Fuel Standard (RFS)*. For more background on biomass definitions, see CRS Report R40529, *Biomass: Comparison of Definitions in Legislation*.

In: Agricultural Conservation: Programs and Issues ISBN: 978-1-61761-288-6
Editors: Steven N. Murphy

Chapter 3

U.S. DEPARTMENT OF AGRICULTURE: IMPROVED MANAGEMENT CONTROLS CAN ENHANCE EFFECTIVENESS OF KEY CONSERVATION PROGRAMS

United States Government Accountability Office

WHY GAO DID THIS STUDY

The U.S. Department of Agriculture (USDA) administers conservation programs, such as the Conservation Stewardship Program (CSP, formerly the Conservation Security Program) and the Environmental Quality Incentives Program (EQIP), to help farmers reduce soil erosion, enhance water supply and quality, and increase wildlife habitat, among other things.

This testimony is based on GAO reports on CSP and EQIP, each issued in 2006, and a 2008 report on farm program payments. It discusses (1) the potential for duplicate payments between CSP and other conservation programs, (2) USDA's process for allocating EQIP funds to the states to optimize environmental benefits, and (3) USDA's management controls over farm program payments.

What GAO Recommends

Among other things, GAO recommended that USDA (1) develop a comprehensive process to preclude and identify duplicate payments between CSP and other conservation programs, (2) take steps to improve the EQIP general financial assistance formula, and (3) work with the Internal Revenue Service (IRS) to develop a method for determining whether all recipients of farm program payments meet income eligibility requirements. USDA agreed with these recommendations and has taken actions to implement them, but GAO has not assessed the effectiveness of these actions.

What GAO Found

While legislative and regulatory measures are in place to reduce the possibility of duplicate payments, the potential still exists because CSP and other USDA conservation programs may be used to finance similar conservation activities. GAO previously reported that USDA did not have a comprehensive process to preclude or identify such duplicate payments, and GAO found a number of instances of duplicate payments. USDA was unaware of this duplication. However, USDA has since updated its contracting software to identify potential duplication and issued written guidance to its field offices outlining measures to preclude duplicate payments. As a result, USDA said that it has identified about 760 examples of potential or actual duplicate payments since fiscal year 2004 totaling about $1 million, and has taken action to preclude or recover these payments, as appropriate.

GAO previously reported that USDA's process for allocating EQIP funds was not clearly linked to the program's purpose of optimizing environmental benefits. Therefore, USDA may not have directed funds to states with the most significant environmental concerns arising from agricultural production. To allocate most EQIP funds, USDA uses a general financial assistance formula that consists of 31 factors and weights. However, USDA did not have a documented rationale for how each factor contributes to accomplishing the program's purpose; some of the formula's data was questionable or outdated; and the funding allocation process was not linked to USDA's long-term performance measures. For fiscal year 2009, USDA has issued updated guidance for this formula that appears to address a number of these elements.

GAO reported that USDA does not have adequate management controls in place to verify that farm program payments, including those for conservation programs, are made only to individuals who do not exceed income eligibility caps. As a result, USDA cannot be assured that millions of dollars in farm payments are proper. GAO found that $49.4 million in farm payments were made to about 2,700 potentially ineligible individuals between fiscal years 2003 and 2006. About 6 percent of this amount was for EQIP payments; 29 percent was for the Conservation Reserve Program, a program that pays farmers to retire environmentally-sensitive cropland. The need for management controls will remain critical, since recent legislation lowered the income eligibility caps and makes the number of individuals whose income exceeds these caps likely to rise. In March 2009, USDA announced that it has begun working with IRS to ensure that high-income individuals and entities who request farm payments meet income limits as set forth in law, and that once this verification system is fully operational, it should identify inappropriate payments before they are disbursed. As GAO has previously reported, ensuring the integrity and equity of farm programs is a key area needing enhanced congressional oversight. Such oversight can help ensure that conservation programs benefit the agricultural sector as intended and protect rural areas from land degradation, diminished water and air quality, and loss of wildlife habitat.

Mr. Chairman and Members of the Subcommittee:

I am pleased to be here today to discuss our work on the U.S. Department of Agriculture's (USDA) management of its conservation programs designed to help farmers be better stewards of our natural resources. Under these programs, primarily the Conservation Stewardship Program (CSP, formerly the Conservation Security Program) and the Environmental Quality Incentives Program (EQIP), USDA and producers (farmers and ranchers) enter into contracts to implement practices to reduce soil erosion, enhance water supply and quality, and increase wildlife habitat, among other things. These conservation programs are administered by USDA's Natural Resources Conservation Service (NRCS). Another USDA agency, the Farm Service Agency (FSA), is responsible for ensuring that only individuals who meet certain eligibility criteria receive federal farm program payments, including payments for many conservation programs.

As you know, farmers and ranchers own or manage about 940 million acres, or about half of the continental United States' land area, and thus they are among the most important stewards of our soil, water, and wildlife habitat.

USDA's conservation programs, which provide billions of dollars in assistance each year, are a key resource in promoting this environmental stewardship. Therefore, it is essential that they be managed effectively and efficiently and that they be adequately overseen to assure that payments are provided only to eligible individuals. We are eager to assist the 111th Congress in meeting its oversight agenda. To that end, we have recommended that ensuring the integrity and equity of the farm programs is a key area needing congressional oversight.[1]

My testimony today is based on our reports on CSP, EQIP, and federal farm program payments.[2] I will focus on three primary issues discussed in these reports: (1) the potential for duplicate payments under CSP and other USDA conservation programs for similar conservation activities, (2) NRCS's process for allocating EQIP funds to the states to optimize environmental benefits, and (3) FSA's efforts to ensure the integrity of farm program payments, including payments for conservation. To perform this work, we reviewed relevant statutory provisions, NRCS, FSA, and other USDA regulations, program documentation, guidelines for implementing EQIP and CSP, and guidance for making farm program payments. We also analyzed data on farm program payments, producer income, and funding allocated to the states under EQIP and to priority watersheds under the Conservation Security Program. In addition, we spoke with officials at NRCS, FSA, other USDA offices, and the Internal Revenue Service (IRS). We conducted our work in accordance with generally accepted government auditing standards.

In summary, USDA has taken a number of actions to address our recommendations to improve its management of these conservation programs and the integrity of farm program payments. Specifically:

- Regarding CSP, we reported that duplicate payments had occurred despite legislative and regulatory measures that were to reduce the potential for duplication between CSP and other programs. We recommended that NRCS develop a process to preclude further duplicate payments as well as to identify and recover past duplicate payments. In response, NRCS updated its contracting software to identify potential duplication and issued written guidance to its field offices in October 2006 outlining measures to preclude duplicate payments. As a result, NRCS reportedly has identified 760 examples of potential or actual duplicate payments since fiscal year 2004 totaling nearly $1 million, and has taken action to preclude or recover these payments, as appropriate.

- Regarding EQIP, we reported that NRCS's formula for allocating financial assistance, which accounts for most of the funding provided to the states, does not link to the program's purpose of optimizing environmental benefits. We recommended that NRCS ensure that the rationale for the formula's factors and weights used to determine the state allocations is documented and linked to program priorities, and that data sources used in the formula are accurate and current. We also recommended that NRCS use information from long-term performance measures to further revise the formula to ensure funds are directed to areas of highest priority. In response, in January 2009, NRCS issued updated guidance for its EQIP funding allocation formula that appears to address a number of the elements raised in our recommendation.
- Regarding the integrity of farm program payments, we reported that USDA cannot be certain that millions of dollars in farm program payments, including conservation payments, it made are proper because it does not have management controls to verify that payments are made only to individuals who did not exceed income eligibility caps. We recommended that FSA work with IRS to develop a method for determining whether all recipients of farm payments meet income eligibility criteria. In response, USDA announced last week that it has begun working with IRS to ensure that high-income individuals and entities who request farm program payments meet income limits as set forth in law. According to USDA, once this verification system is fully operational, it should identify inappropriate payments before they are disbursed.

While these are positive steps, we have not evaluated their effectiveness. In the latter two cases, the agency actions to implement our recommendations are so recent that there is little or no basis yet to do this evaluation.

LEGISLATIVE AND REGULATORY MEASURES REDUCE THE POTENTIAL FOR DUPLICATION BETWEEN CSP AND OTHER PROGRAMS, BUT DUPLICATE PAYMENTS HAVE OCCURRED

EQIP provides assistance to farmers and ranchers to take new actions aimed at addressing identified conservation problems. CSP rewards farmers and ranchers who already meet very high standards of conservation and environmental management in their operations. Farm bill provisions and NRCS regulations are designed to reduce the potential for duplication between CSP and other USDA conservation programs, such as EQIP. For example, the Farm Security and Rural Investment Act of 2002 (2002 farm bill) and the Food, Conservation, and Energy Act of 2008 (2008 farm bill)[3]

- provide that CSP may reward producers for maintaining conservation practices that they have already undertaken, whereas other programs generally provide assistance to encourage producers to take new actions to address conservation problems on working lands or to idle or retire environmentally sensitive land from agricultural production; and
- explicitly prohibit (1) duplicate payments under CSP and other conservation programs for the same practice on the same land and (2) CSP payments for certain activities that can be funded under other conservation programs, such as the construction or maintenance of animal waste storage or treatment facilities.

  USDA has also issued CSP regulations that can prevent duplicate payments between CSP and other conservation programs. For example, the regulations
- establish higher minimum eligibility standards for CSP than for other programs, which help to differentiate the applicant pool for CSP from the potential applicants for these other programs; and
- encourage CSP participants to implement conservation actions, known as enhancements, to achieve a level of treatment that generally exceeds the level required by other USDA conservation programs.

Despite these legislative and regulatory measures, we reported in 2006 that the potential for duplicate payments still existed because of similarities in conservation actions financed through CSP and other programs. At that time, we found that duplicate payments had occurred. Our analysis of fiscal year 2004 payments data showed 72 producers who received payments under CSP and EQIP that appeared to be for similar conservation actions. Of these, we examined 11 cases in detail and found duplicate payments had occurred 8 times. For example, four of these duplicate payments were made to producers who received a CSP enhancement payment and an EQIP payment for conservation actions that appeared to be similar. In one of these cases, a producer received a CSP pest management enhancement payment of $9,160 and an EQIP payment of $795 on the same parcel of land for the same conservation action—conservation crop rotation.

NRCS state officials agreed that the payments made in these four cases were duplicates. They stated that they were unaware that such duplication was occurring and that they would inform their district offices of it. At the time of our report, NRCS headquarters officials stated that the agency lacked a comprehensive process to either preclude duplicate payments or identify them after a contract has been awarded. Instead, these officials said, as a guard against potential duplication, NRCS relied on the institutional knowledge of its field staff and the records they keep.

NRCS has the authority to recover duplicate payments. Under a CSP contract, as required in the 2002 and 2008 farm bills, a producer agrees that if the producer violates any term or condition of the contract, the producer is to refund payments and forfeit all rights to receive payments or is to refund or accept adjustments to payments, depending on whether the Secretary of Agriculture determines that termination of the contract and return of payments is or is not warranted, respectively.

Duplicate payments reduce program effectiveness and, because of limited funding, may result in some producers not receiving program benefits for which they are otherwise eligible. For these reasons, we recommended that the Secretary of Agriculture direct the Chief of NRCS to develop processes to review (1) CSP contract applications to ensure that CSP payments, if awarded, would not duplicate payments made by other USDA conservation programs and (2) existing CSP contracts to identify cases where CSP payments duplicate payments made under other programs and take action to recover appropriate amounts and to ensure that these duplicate payments are not repeated in fiscal year 2006 and beyond.

Regarding the first recommendation, in July 2006, NRCS said it had created an automated system within its contracting software to conduct a comparison between new CSP applications and existing contracts for other conservation programs to reveal potential duplication. In addition, in October

2006, NRCS issued a national bulletin to its field staff describing measures needed to preclude duplicate payments. According to the bulletin, NRCS conducted a comparison between existing contracts for several conservation programs, including EQIP, and fiscal year 2006 CSP applications to identify potential duplication. This comparison found 81 potential duplicate payments for conservation practices. NRCS said it adjusted the CSP applications to prevent these duplicate payments. Furthermore, NRCS indicated that starting with the fiscal year 2006 CSP sign-up, it would require applicants to complete a form that asks an applicant to identify any payments the applicant receives under another conservation program on any of the land being offered for enrollment in CSP. While these actions are positive steps, we have not assessed their effectiveness.

Regarding the second recommendation, NRCS indicated it would use its contracting software to compare existing CSP contracts with existing contracts for EQIP and other conservation programs. Specifically, according to NRCS's national bulletin, its field offices are to compare CSP contract enhancement activities with the practices financed under other conservation program contracts to determine whether duplicate payments are planned in fiscal year 2007 and beyond, or if duplicate payments occurred during fiscal years 2004 through 2006. NRCS said that all identified duplicate payments would be dealt with according to the NRCS contracting manual. According to NRCS officials, the agency did not have a CSP sign-up in 2007, so there were no new applications that year. In 2008, NRCS received about 2,300 CSP applications, but agency officials said they did not have information on potential duplicate payments. For 2004 to 2006, NRCS officials said the agency found 371 duplicate payments between CSP and EQIP totaling about $420,000. These officials did not have information on the amount of these payments recovered, but noted that they represented less than 1 percent of total CSP payments made during these years. Furthermore, NRCS officials stated the agency found 389 scheduled payments totaling about $520,000 under these programs that would have been duplicates. NRCS was able to preclude these payments from being made.

NRCS's Process for Allocating EQIP Funds to the States Does Not Link to the Program's Purpose of Optimizing Environmental Benefits

In 2006, we reported that NRCS's process for providing EQIP funds to the states is not clearly linked to the program's purpose of optimizing environmental benefits. Specifically, we found that NRCS's general financial assistance formula, which accounts for approximately two-thirds of funding provided to the states, did not have a documented rationale for each of the formula's factors and weights, which are used to determine the allocation of funds to the states to address environmental issues. In addition, the formula sometimes relied on questionable and outdated data. As a result, NRCS may not have been directing EQIP funds to states with the most significant environmental concerns arising from agricultural production.

More specifically, in fiscal year 2006, approximately 65 percent of EQIP funds were allocated using a general financial assistance formula. This formula contained 31 factors related to the availability of natural resources and the presence of environmental concerns, such as acres of wetlands and at-risk species habitat, pesticide and nitrogen runoff, and the ratio of commercial fertilizers to cropland. NRCS assigns each of the formula's factors a weight. Factors with the highest weights included acres of highly erodible cropland, acres of fair and poor rangeland, the quantity of livestock, and the quantity of animal waste generated.

At the time of our report, NRCS had periodically modified factors and weights to emphasize different national priorities, such as in fiscal year 2004, following the passage of the 2002 farm bill. However, NRCS had not documented the basis for its decisions on the formula factors and weights or explained how they achieve the program's purpose of optimizing environmental benefits. Thus, it was not always clear whether the formula's factors and weights directed funds to the states as effectively as possible.

Small differences in the weights can shift the amount of financial assistance directed at a particular concern. For example, in 2006, if the weight of any of the 31 factors had increased by 1 percent, $6.5 million would have been shifted at the expense of one or more other factors. The potential for the weights to significantly affect the amount of funding a state receives underscores the importance of having a well-founded rationale for assigning them.

We also reported that weaknesses in the financial assistance formula were compounded by NRCS's use of questionable and outdated data. First, 5 of the 29 data sources in the financial assistance formula were used more than once for separate factors. Using the same data for multiple factors may result in more emphasis being placed on certain environmental concerns than intended. Second, NRCS could not confirm the source of data used in 10 factors in the formula; as such, we could not determine the accuracy of the data, verify how NRCS generated the data, or fully understand the basis on which the agency allocates funding. Third, NRCS did not use the most current data for six factors in the formula.

Finally, we reported that NRCS had begun to develop more long-term, outcome-oriented performance measures to assess changes to the environment resulting from EQIP practices as part of its 2005 strategic planning effort. These measures included such things as reducing sediment runoff from farms, improving soil conditions on working cropland, and increasing water conservation. NRCS also included proposed targets for each measure to be achieved by 2010, such as reducing sediment runoff by 18.5 million tons annually. At the time of our report, NRCS told us it had developed baselines for these performance measures, and planned to assess and report on them once computer models and other data collection methods that estimate environmental change were completed.

Although we did not assess the comprehensiveness of the EQIP performance measures, the additional information they provide about the results of EQIP outcomes should allow NRCS to better gauge program performance. As a next step, such information could also help the agency refine its process for allocating funds to the states through its general financial assistance formula by directing funds toward practices that address unrealized performance targets and areas of the country that need the most improvement. The Chief of NRCS's Environmental Improvement Programs Branch agreed that information about program performance might eventually be linked to the EQIP funding allocation process. However, at the time of our report, the agency did not have plans to make this linkage.

Because of our concerns about the general financial assistance formula, we recommended that NRCS ensure its rationale for the factors and weights was documented and addressed program priorities, and the data sources used in the formula were accurate and current. We also recommended that the Secretary of Agriculture direct NRCS to continue to analyze current and newly developed long-term performance measures for EQIP and use this information to make further revisions to the financial assistance formula to ensure funds are directed to areas of highest priority.

Since our report, NRCS has made progress in implementing our recommendations by modifying its financial assistance formula for the fiscal year 2009 EQIP state allocation. In 2007, an outside consultant hired by NRCS concluded that NRCS should take a number of steps to improve its conservation program formulas, including improving their analytical soundness, making the process more transparent, and integrating performance information into the formulas. NRCS reviewed the EQIP formula and made changes prior to its 2009 allocation, including modifying the factors and weights, and updating some data sources. NRCS also described how factors in the formula relate to a number of EQIP and NRCS performance measures. While NRCS's actions are positive steps, we have not assessed whether they fully address our recommendations.

ADDITIONAL USDA MANAGEMENT CONTROLS COULD PROVIDE MORE ASSURANCE OF CONSERVATION PROGRAM INTEGRITY

Additional management controls by USDA's FSA could provide more assurance of the conservation programs' integrity by ensuring conservation payments are awarded only to individuals who meet income eligibility requirements.[4] In October 2008 we reported that USDA cannot be certain that millions of dollars in farm program payments it made are proper, because it does not have management controls, such as reviewing an appropriate sample of recipients' tax returns, to verify that payments were made only to individuals who did not exceed the income eligibility caps. We determined that $49.4 million in farm payments were made to about 2,700 potentially ineligible individuals between fiscal year 2003 and fiscal year 2006. These recipients included a founder and former executive of an insurance company, an individual with ownership interest in a professional sports franchise, a top executive of a major financial services company, a former executive of a technology company, and individuals residing outside the United States.

As shown in figure 1, about 6 percent of the $49.4 million was for EQIP payments and 29 percent was for the Conservation Reserve Program. Payments made under the "other programs" category included payments made for other NRCS conservation programs, such as CSP, the Grassland Reserve

Program, Wetlands Reserve Program, and Wildlife Habitat Incentives Program.

According to FSA officials, a number of factors—such as resource constraints that hamper its ability to examine complex tax and financial information and lack of authority to access and use IRS tax filer data for such purposes—contribute to its inability to verify that each individual who received farm program payments was eligible. We also found, however, that the sample FSA draws to check recipient eligibility does not test for income eligibility; instead, FSA reviews compliance with eligibility requirements other than income, such as how much a farming operation received in farm program payments in the previous year and whether it experienced a change in ownership. FSA therefore cannot ensure that only individuals who meet the income eligibility caps are receiving farm payments.

Without better management controls, USDA cannot be assured that millions of dollars in farm program payments, including conservation payments, are proper. This need for management controls will remain critical, since the 2008 farm bill lowered the income eligibility caps. This change makes the number of individuals whose adjusted gross income exceeds the caps likely to rise, which increases the risk that USDA could make improper payments to more individuals.

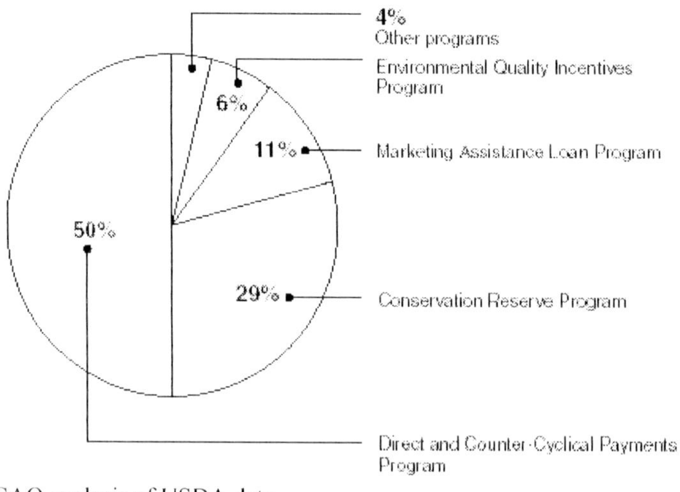

Source: GAO analysis of USDA data.

Figure 1. Percentage of $49.4 Million Paid to Potentially Ineligible Individuals, by Program, Fiscal Years 2003 through 2006

To ensure greater program integrity, we recommended that the Secretary of Agriculture direct FSA to work with IRS to develop a method for determining whether all recipients of farm program payments meet income eligibility requirements, and, if the Secretary finds that USDA does not have authority to obtain information from IRS, request the authority it would need from Congress. USDA agreed with our recommendations and, in a March 19, 2009, news release, the agency announced that it would work with IRS to ensure that high-income individuals and entities who request USDA payments meet income limits set forth in the 2008 farm bill. Specifically, in order to be eligible for USDA payments all recipients will be required to sign a separate form that grants IRS authority to provide income information to USDA for verification purposes. According to USDA, once this verification system is fully operational, it should identify inappropriate payments before they are disbursed.

CONCLUSIONS

In conclusion, USDA conservation programs can play an invaluable role in encouraging farmers and ranchers to act as stewards of the nation's natural resources. However, the weaknesses we previously identified in the management of CSP and EQIP funds, as well as our concerns with controls related to farm program payments more generally, could undermine the effectiveness of USDA conservation programs. On a positive note, in response to our recommendations, USDA has taken a number of promising actions to eliminate duplicate payments between CSP and other programs, refine the EQIP allocation formula by updating its factors, weights, and data sources and, in some cases, identifying how the factors relate to long-term performance measures, and strengthen management controls over farm program payments. While these actions are positive, continued oversight of these programs, such as today's hearing, helps ensure funds are spent as economically, efficiently, and effectively as possible and benefit the agricultural sector as intended. Such oversight is especially critical in light of the nation's current deficit and growing long-term fiscal challenges.

Mr. Chairman, this concludes my prepared statement. I would be pleased to respond to any questions that you or other Members of the Subcommittee may have.

CONSERVATION SECURITY PROGRAM: DESPITE COST CONTROLS, IMPROVED USDA MANAGEMENT IS NEEDED TO ENSURE PROPER PAYMENTS AND REDUCE DUPLICATION WITH OTHER PROGRAMS

WHY GAO DID THIS STUDY

The Conservation Security Program (CSP)—called for in the 2002 farm bill and administered by the U.S. Department of Agriculture's (USDA) Natural Resources Conservation Service (NRCS)—provides financial assistance to producers to reward past conservation actions and to encourage further conservation stewardship. CSP payments may be made for structural or land management practices, such as strip cropping to reduce erosion. CSP has raised concerns among some stakeholders because CSP cost estimates generally have increased since the 2002 farm bill's enactment. For example, the Congressional Budget Office's estimate increased from $2 billion in 2002 to $8.9 billion in 2004.

GAO determined (1) why CSP cost estimates generally increased; (2) what authority USDA has to control costs and what cost control measures exist; and (3) what measures exist to prevent duplication between CSP and other USDA conservation programs and what duplication, if any, has occurred.

Source: Photo courtesy of USDA NRCS.

Note: Strip cropping means growing row crops, forages, or small grains in equal width strips.

Strip Cropping to Reduce Soil Erosion

WHAT GAO RECOMMENDS

GAO recommends, in part, that NRCS review its state offices' wildlife habitat assessment criteria and develop a process to preclude and identify duplicate payments. NRCS generally agreed with GAO's findings and recommendations.

WHAT GAO FOUND

Various factors explain why estimates of CSP costs generally increased since the 2002 farm bill's enactment. Of most importance, little information was available regarding how this program would be implemented at the time of its inception in 2002. As more information became available, cost estimates rose. In addition, the time frames on which the estimates were based changed. While the initial estimates covered years in which the program was expected to be nonoperational or minimally operational, subsequent estimates did not include these years.

The farm bill provides USDA general authority to control CSP costs, including authority to establish criteria that enable it to control program participation and payments and, therefore, CSP costs. For example, NRCS restricts participation by limiting program enrollment each year to producers in specified, priority watersheds. NRCS also has established certain CSP payment limits at levels below the maximum allowed by the statute. However, efforts to control CSP spending could be improved by addressing weaknesses in internal controls and inconsistencies in the wildlife habitat assessment criteria that NRCS state offices use, in part, to determine producer eligibility for the highest CSP payment level. Inconsistencies in these criteria also may reduce CSP's conservation benefits.

The farm bill prohibits duplicate payments for the same practice on the same land made through CSP and another USDA conservation program. Various other farm bill provisions also reduce the potential for duplication. For example, as called for under the farm bill, CSP may reward producers for conservation actions they have already taken, whereas other programs generally provide assistance to encourage new actions or to idle or retire environmentally sensitive land from production. In addition, CSP regulations establish higher minimum eligibility requirements for CSP than for other programs. However, despite these legislative and regulatory provisions, the possibility that producers can receive duplicate payments remains because of similarities in the conservation actions financed through these programs. In addition, NRCS does not have a comprehensive process to preclude or identify such duplicate payments. In reviewing NRCS's payments data, GAO found a number of examples of duplicate payments.

Agricultural Conservation: USDA Should Improve Its Process for Allocating Funds to States for the Environmental Quality Incentives Program

Why GAO Did This Study

The Environmental Quality Incentives Program (EQIP) assists agricultural producers who install conservation practices, such as planting vegetation along streams and installing waste storage facilities, to address impairments to water, air, and soil caused by agriculture or to conserve water. EQIP is a voluntary program managed by the U.S. Department of Agriculture's (USDA) Natural Resources Conservation Service (NRCS). NRCS allocates about $1 billion in financial and technical assistance funds to states annually. About $650 million of the funds are allocated through a general financial assistance formula.

As requested, GAO reviewed whether USDA's process for allocating EQIP funds to states is consistent with the program's purposes and whether USDA has developed outcome-based measures to monitor program performance. To address these issues, GAO, in part, examined the factors and weights in the general financial assistance formula.

What GAO Recommends

GAO recommends, among other things, that NRCS document its rationale for the factors and weights in its general financial assistance formula and use current and accurate data. USDA agreed with GAO that the formula needed review. USDA did not agree with GAO's view that NRCS's funding process does not clearly link to EQIP's purpose of optimizing environmental benefits.

It believes that the funding process clearly links to EQIP's purpose, but it has not documented the link.

WHAT GAO FOUND

NRCS's process for providing EQIP funds to states is not clearly linked to the program's purpose of optimizing environmental benefits; as such, NRCS may not be directing funds to states with the most significant environmental concerns arising from agricultural production. To allocate most EQIP funds, NRCS uses a general financial assistance formula that consists of 31 factors, including such measures as acres of cropland, miles of impaired rivers and streams, and acres of specialty cropland. However, this formula has several weaknesses. In particular, while the 31 factors in the financial assistance formula and the weights associated with each factor give the formula an appearance of precision, NRCS does not have a specific, documented rationale for (1) why it included each factor in the formula, (2) how it assigns and adjusts the weight for each factor, and (3) how each factor contributes to accomplishing the program's purpose of optimizing environmental benefits. Factors and weights are important because a small adjustment can shift the amount of funding allocated to each state on the basis of that factor and, ultimately, the amount of money each state receives. For example, in 2006, a 1 percent increase in the weight of any factor would have resulted in $6.5 million more allocated on the basis of that factor and a reduction of 1 percent in money allocated for other factors. In addition to weaknesses in documenting the design of the formula, some data NRCS uses in the formula to make financial decisions are questionable or outdated. For example, the formula does not use the most recent data available for 6 of the 31 factors, including commercial fertilizers applied to cropland. As a result, any recent changes in a state's agricultural or environmental status are not reflected in the funding for these factors. During the course of GAO's review, NRCS announced plans to reassess its EQIP financial assistance formula.

NRCS recently developed a set of long-term, outcome-based performance measures to assess changes to the environment resulting from EQIP practices. The agency is also in the process of developing computer models and other data collection methods that will allow it to assess these measures. Thus, over time, NRCS should ultimately have more complete information on which to gauge program performance and better direct EQIP funds to areas of the country that need the most improvement.

Federal Farm Programs: USDA Needs to Strengthen Controls to Prevent Payments to Individuals Who Exceed Income Eligibility Limits

Why GAO Did This Study

Farmers receive about $16 billion annually in federal farm program payments. These payments go to about 2 million recipients, both individuals and entities. GAO previously has reported that the U.S. Department of Agriculture (USDA) did not consistently ensure that these payments went only to those who meet eligibility requirements.

GAO was asked to evaluate (1) how effectively USDA implemented 2002 Farm Bill provisions prohibiting payments to individuals or entities whose income exceeded $2.5 million and who derived less than 75 percent of that income from farming, ranching, or forestry operations, (2) the potential impact of the 2008 Farm Bill's income eligibility provisions on individuals who receive farm payments, and (3) the distribution of income of these individuals compared with all 2006 tax filers. GAO compared USDA data on individuals receiving payments with the latest available Internal Revenue Service (IRS) data on these individuals.

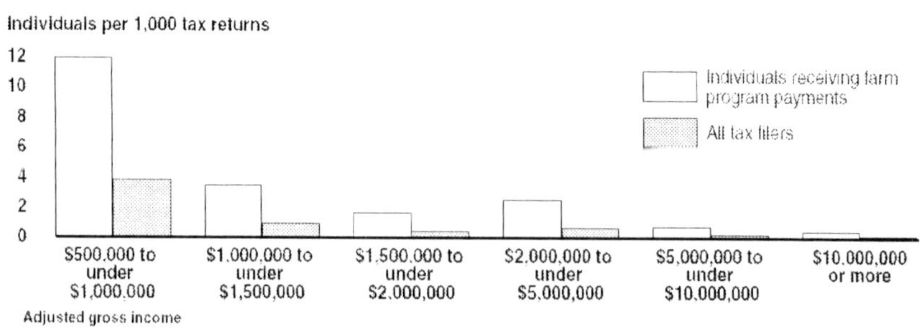

Source: GAO analysis of USDA and IRS data.

Distribution of Income of Individuals Receiving Farm Program Payments and All Tax Filers, 2006

WHAT GAO RECOMMENDS

GAO recommends that USDA work with IRS to develop a system for verifying the income eligibility for all recipients of farm program payments. If USDA determines that it needs authority to work with IRS, it should seek this authority from Congress, as appropriate. In commenting on a draft of this report, USDA agreed with these recommendations but disputed some of the findings. GAO believes that the report is fair and accurate.

WHAT GAO FOUND

USDA does not have management controls, such as reviewing an appropriate sample of recipients' tax returns, to verify that payments are made only to individuals who do not exceed income eligibility caps and therefore cannot be assured that millions of dollars in farm program payments it made are proper. GAO found that of the 1.8 million individuals receiving farm payments from 2003 through 2006, 2,702 had an average adjusted gross income (AGI) that exceeded $2.5 million and derived less than 75 percent of their income from farming, ranching, or forestry operations, thereby making them potentially ineligible for farm payments. Nevertheless, USDA paid over $49 million to these individuals. According to USDA officials, a number of factors—such as resource constraints that hamper its ability to examine complex tax and financial information as well as a lack of authority to obtain and use IRS tax filer data for such purposes—contribute to the department's inability to verify that each individual who receives farm program payments complies with income eligibility provisions. However, USDA does not routinely sample individuals receiving farm payments to test for income eligibility; instead, its annual sample selected for review is based primarily on compliance with eligibility requirements other than income. The 2008 Farm Bill directs USDA to use statistical methods to target those individuals most likely to exceed income eligibility caps.

The 2008 Farm Bill will increase the number of individuals likely to exceed the income eligibility caps. That is, with lower income eligibility caps under the 2008 Farm Bill, the number of individuals whose AGI exceeds the caps will rise, increasing the risk that USDA will make improper payments to more individuals. For example, had the new Farm Bill been in effect in 2006, as many as 23,506 individuals who received farm program payments would likely have been ineligible for crop subsidy and disaster assistance payments totaling as much as $90 million.

Compared with all tax filers, individuals who participated in farm programs in 2006 are more likely to have higher incomes. For example, as shown in the figure below, 12 of every 1,000 individuals receiving farm program payments reported AGI between $500,000 and $1 million compared with about 4 of all tax filers who reported income at this level.

End Notes

[1] GAO, *Suggested Areas for Oversight for the 110th Congress*, GAO-07-235R (Washington, D.C.: Nov. 17, 2006).

[2] GAO, *Conservation Security Program: Despite Cost Controls, Improved USDA Management Is Needed to Ensure Proper Payments and Reduce Duplication with Other Programs*, GAO-06-312 (Washington, D.C.: Apr. 28, 2006). GAO, *Agricultural Conservation: USDA Should Improve Its Process for Allocating Funds to States for the Environmental Quality Incentives Program*, GAO-06-969 (Washington, D.C.: Sept. 22, 2006). GAO, *Federal Farm Programs: USDA Needs to Strengthen Controls to Prevent Payments to Individuals Who Exceed Income Eligibility Limits*, GAO-09-67 (Washington, D.C.: Oct. 24, 2008). Copies of the Highlights pages for these reports are attached to this statement.

[3] The Conservation Security Program was originally authorized in the 2002 farm bill and included measures to reduce the potential for duplication with other USDA conservation programs. Similar measures are also included in the Conservation Stewardship Program authorized in the 2008 farm bill.

[4] Although these limits changed in the 2008 Farm Bill, under the 2002 Farm Bill, an individual or entity with an average adjusted gross income (AGI) of over $2.5 million, over the previous 3 tax years immediately preceding the applicable crop year, was ineligible for farm program payments unless at least 75 percent or more of the average AGI was farm income, defined as income from farming, ranching, or forestry operations. The AGI provision of the 2002 Farm Bill covered crop years 2003 through 2008 and applied to most farm program payments, including those for crop subsidy payments (e.g., fixed payments based on historical production, known as direct payments, and price support payments), conservation practices, and disasters.

In: Agricultural Conservation: Programs and Issues ISBN: 978-1-61761-288-6
Editors: Steven N. Murphy

Chapter 4

STATEMENT OF ROBERT STEPHENSON, ACTING DEPUTY ADMINISTRATOR FOR FIELD OPERATIONS, FARM SERVICE AGENCY, U.S. DEPARTMENT OF AGRICULTURE, AND DAVID WHITE, CHIEF, NATURAL RESOURCES CONSERVATION SERVICE, U.S. DEPT. OF AGRICULTURE, BEFORE THE SUBCOMMITTEE ON CONSERVATION, CREDIT, ENERGY, AND RESEARCH

Robert Stephenson and David White

Mr. Chairman and members of the subcommittee, we appreciate the opportunity to review conservation programs delivered by the U.S. Department of Agriculture (USDA). We are pleased to share our experiences in implementing the Conservation Title. We will also offer our observations on the changing business environment in which programs operate, the working relationships with our USDA conservation partners, and the opportunities and challenges we face in implementing the 2008 Farm Bill.

Farm Service Agency

Background and Programs

The Farm Service Agency (FSA) delivers conservation, commodity, credit, and emergency programs. Program level funding varies depending upon market and weather conditions and new legislation. For fiscal years (FYs) 2007 and 2008, the program level was $30.8 billion and $25.0 billion, respectively. We estimate the level to be $23.7 billion for FY 2009. FSA has a staffing level of just under 14,700 staff years and an annual salaries and expenses budget of about $1.5 billion.

FSA's conservation programs include the Conservation Reserve Program (CRP), Emergency Conservation Program (ECP), Grass Roots Source Water Program (Source Water), Voluntary Public Access and Habitat Incentive Program (Public Access), and the Emergency Forestry Restoration Program. FSA also shares program delivery with the Natural Resources Conservation Service (NRCS) of the Grassland Reserve Program.

Implementation Model

Most FSA programs are delivered through a network of State and county offices that are located in over 2,200 rural counties. Other programs, such as Source Water, are implemented through the National Rural Water Association and Public Access is implemented as grants to State and Tribal governments.

At the contract level, under CRP, FSA assists farmers and ranchers with enrolling land, ensuring compliance with program goals and requirements, managing the contract, making payments, and obtaining technical assistance which is generally provided by NRCS or local conservation districts. In some cases, non-Government providers may also offer technical assistance which includes practice eligibility determinations and conservation plan development.

In delivering its conservation programs, FSA has entered into agreements with some of its partners to provide technical support. Chief among those agreements is FSA's relationship with NRCS. Since the 1930's, FSA and NRCS employees have worked closely together to assist farmers and ranchers in conserving and improving our Nation's natural resources.

The NRCS role included developing technical standards and providing technical assistance. Over time, NRCS' role has expanded in the area of program delivery as this Committee has added a number of important conservation programs to the NRCS portfolio including the Environmental Quality Incentives Program (EQIP), Conservation Security Program, and Wetlands Reserve Program (WRP).

FSA's agreement with NRCS for CRP includes providing technical assistance. Other Government partners include USDA's Forest Service (FS) and Cooperative State Research, Education, and Extension Service; State forestry agencies, and local soil and water conservation districts.

FSA, NRCS, and FS have a long history of delivering conservation programs to farm and ranch community. Since the Dust Bowl days of the 1930's, FSA and NRCS have been partners in delivering conservation programs' financial and technical assistance. The success of our efforts is seen across the landscape in windbreaks, waterways, filterstrips, and wetlands implemented through programs such as conservation compliance, ACP, EQIP and CRP.

Both Agencies are committed to the delivery of conservation program that will "get conservation on the ground" in an efficient and effective manner. We take our fiduciary responsibilities seriously and want to be accountable to the public for our performance. These common goals require the Agencies to work together and with our partners.

At the National level, the Agencies jointly work in the development of program policies such as CRP. The Agencies meet on a regular basis to discuss resource allocation issues and ways to improve program performance. In the case of CRP, FSA administers the program but utilizes the strength of Agencies such as NRCS and FS for providing technical assistance.

NRCS and FS are recognized as leaders in developing conservation practice technical standards and conservation plans and providing conservation technical assistance. Also, soil surveys and natural resource and forest inventories are critical components of designing effective conservation programs.

FSA has been delivering conservation programs since the 1930's. Since the 1980's, FSA and its partners, including NRCS, transformed the CRP program from primarily an erosion control program to a multi-dimensional conservation program that now addresses water quality, wildlife, water quantity, threatened and endangered species, and carbon sequestration issues.

2008 Farm Bill Implementation

The 2008 Farm Bill responded to a broad range of ongoing conservation challenges including soil erosion, wetlands conservation, water quality, wildlife habitat, and potential markets for sequestered carbon and other environmental services.

The 2008 Farm Bill re-authorized CRP and Source Water and authorized, for the first time, Public Access and the Emergency Forestry Restoration Program.

The CRP-related provisions will be implemented in two parts. We are working diligently on Part 1, which includes the Farmable Wetland Program (i.e., aquaculture restoration, constructed wetlands, flooded prairie wetlands, and wetland restoration), tree thinning, and the conservation exception under the new Average Adjusted Gross Income provisions.

The other CRP-related provisions of the 2008 Farm Bill which includes cropping history requirements, transition payment to beginning and socially disadvantaged farmers and ranchers, and routine grazing are scheduled to be implemented after completion of an Environmental Impact Statement.

Public Access provides grants to State Governments and Tribes to expand public access opportunities on private land and is scheduled to be implemented later this year.

The Emergency Forestry Restoration Program will assist in the restoration of forests damaged due to natural disasters including replanting. An appropriation of funds is necessary to implement.

Program Accomplishments

America's farmers and ranchers have made significantly strides to lessen the impact on our Nation's environment over the last 20 years. Under all USDA conservation programs, soil erosion on cropland has been reduced by over 1.2 billion tons per year. As of February 2009, CRP participants have restored more than 2 million acres of wetlands and about 2 million acres of buffers. Land enrolled in CRP will also reduce soil erosion by 400 million tons each year and has the potential to be one of Nation's largest carbon sequestration programs on private lands.

During October 2008, FSA issued over 900,000 checks to CRP participants and most of the participants received their payment with a few days after they were eligible. FSA maintains many of the databases that are essential including Average Adjusted Gross Income, conservation compliance, financial offset. FSA also works extensively with NRCS to integrate our databases to assist them in implementing programs such as Environmental Quality Incentive Program, Grassland Reserve Program, and other programs.

Program Performance – Financial

In an environment of increasing public service demands, scrutiny and decreasing resources, FSA has improved program integrity and fiscal stewardship by enhancing internal controls, transparency, and accountability in USDA's financial management programs. By recognizing that strong internal controls and solid financial management practices are the cornerstones of effective federal stewardship, FSA has focused much of this effort on working to address weaknesses.

By developing and implementing corrective action plans that ensured a correct measurement of improper paperwork and improper payments, FSA was able to reduce its improper payments reported from $2.9 billion (11.2 percent) to $187 million (1.3 percent) between FYs 2006 and 2008. In addition, commitment to continuous improvement to strengthening internal controls and accountability has resulted in the achievement in seven consecutive Commodity Credit Corporation (CCC) unqualified or "clean" financial statement audit opinions, testimony that the CCC's financial statement data is reliable, accurate, and complete.

FSA continues to work on improving our financial controls for our program. From FY 2006 through FY 2008, we conducted reviews under the Improper Payments Information Act (IPIA) to determine the potential extent of improper payments and ways to improve our business process.

These statistical surveys indicated that the error rate for improper payments for CRP was 3.53 percent for FY 2006 which was reduced to 1.25 percent for FY 2008. For CRP and other programs, this reduction was achieved through an aggressive commitment by the Agency which included: (1) direct senior management involvement; (2) agency-wide training; (3) increased accountability at levels; (4) development and use of checklists; (5) enhanced program eligibility verification; (6) elimination of automatic rollover of eligibility determinations; (7) improved documentation control; (8) a comprehensive re-examination of payment files; and (9) increased internal controls and external audits.

Future Outlook

Further improvements in financial integrity are planned. Under CRP, software to record financial obligations at the contract level is scheduled for release within the year.

The recently enacted Stimulus Bill provided $50 million to assist with the stabilization and modernization of FSA's Information Technology systems. This funding will be used to continue essential investments to stabilize the infrastructure and performance of the web-based systems and to initiate the modernization program to provide a modern-day IT system architecture supporting Farm Program delivery and moving away from the 1980's era technologies used today.

We also have ongoing efforts to: (1) improve data quality and develop a data warehouse; (2) improve the governance and the quality of user requirements; and (3) to improve and standardize common business process. These efforts all require significant staff and financial resources.

Geospatial Information Systems (GIS) is an innovative technology that FSA and NRCS have been working with over the last decade to change the way the Agencies manage conservation programs. GIS provides an intuitive solution for managing, visualizing, and understanding land information that enables more efficient management of conservation programs.

FSA and NRCS have acquired and developed a substantial collection of computerized map assets such as soil survey, aerial imagery (NAIP), farm field boundaries (Common Land Unit that describes the agricultural activities nationwide), and others that are used both internal to USDA and are available to the wide range of customers via data centers and data warehouses.

Integration of these powerful resources into everyday business processes is an ongoing challenge to the Agencies but significant progress has been made in laying the foundation for implementing cost-effective and common sense solutions to better support FSA conservation efforts and conservation program delivery. GIS has the capability to support and enable better decision-making and effective solutions to the wide range of conservation issues that FSA faces in the coming years.

While environmental indicators clearly show progress in resource conservation is being made, many challenges remain and new issues continue to emerge. For example, excess nutrients impair water quality in many rivers, streams, and lakes, and hypoxia is a significant problem in the Gulf of Mexico, Chesapeake Bay, and other waters. In addition, conflicts over water availability for agriculture, environmental, and urban use are increasing as water demands increase. As one of the largest water users, agriculture has a vital interest in securing water quality and quantity. Conservation is bringing about important achievements, but more can be done, particularly for wetland and aquatic systems.

In the near term, CRP contracts enrolling about 3.9 million acres are scheduled to expire on September 30, 2009. Taking into account the reduced enrollment authority of 32.0 million acres and ongoing enrollment for continuous signup practices, there is some room under the cap to enroll more acres, though there is insufficient authority to re-enroll all of these acres. The lost conservation benefit could result in increases in water and air pollution and could exacerbate recovery of the Lesser Prairie Chicken in the southern Great Plains.

Natural Resources Conservation Service

Conservation Investments and Trends

Before getting into the operational mechanics of the NRCS conservation programs, I would like to take just a moment to put the federal investment in agricultural conservation programs into perspective. Consider for a moment the following trends in conservation program investments just in the past twelve years:

- In 1996, many of the conservation programs that are so familiar today were just in their infancy. Congress created and authorized EQIP at $200 million per year, but it was regularly limited to nearly $170 million per year.
- In 1996, new programs such as the Farm and Ranch Lands Protection Program (FRPP) and Wildlife Habitat Incentives Program (WHIP) were funded at $35 million and $50 million *total* over the life of that Farm Bill.
- From the 1996 to 2002 Farm Bills, conservation program investments were increased by more than $17 billion over the previous baseline of spending, with programs such as EQIP receiving over a billion in annual spending. FRPP and WHIP greatly expanded in scope and ambitious new programs such as the Conservation Security Program were created.

- The 2008 Farm Bill continued this support with an additional increase of more than $4 billion over the previous baseline.
- Today, NRCS implements more than 20 conservation programs and initiatives, with an annual budget of more than $3 billion.

2008 Accomplishments

The significant investments made by this Subcommittee in Farm Bill conservation programs, combined with the complete range of conservation authorities and initiatives are generating impressive results. USDA appreciates the ongoing support of this Subcommittee to ensure that farmers and ranchers have the financial and technical resources they need to realize their conservation goals. Consider for a moment the conservation accomplishments from last year:

- During FY 2008, NRCS employees helped develop conservation plans covering more than 42 million acres of privately owned farm, ranch, and forestland. We also assisted producers and other land managers to voluntarily implement conservation practices on nearly 50 million acres. These actions on private lands yield public benefits we all enjoy in the form of cleaner and more abundant water, cleaner air, improved wildlife habitat and healthier soils.
- NRCS provided more than $2 billion in financial assistance to landowners and communities to encourage participation in programs such as EQIP, WHIP, CSP, FRPP and others, resulting in tens of thousands of cost share and incentive contracts and easements.
- Volunteers contributed over 810,000 hours to NRCS efforts—valued at over $15 million. The agency also expanded conservation implementation capacity through the certification and re-certification of several hundred Technical Service Providers.
- Beyond delivering planning and technical assistance, NRCS influenced the acceleration and adoption of new technologies, standards and approaches through Conservation Innovation Grants and our National Technology Support Centers.
- The NRCS Snow Survey and Water Supply Forecasting program issued 12,500 water supply forecasts and we mapped or updated soil surveys for over 35 million acres.

Cumulative Results

Looking at the implementation of conservation programs just since the beginning of this decade, NRCS has worked with farmers, ranchers, and landowners to:

- Apply conservation plans and systems on 328 million acres.
- Apply conservation practices through the Environmental Quality Incentives Program (EQIP) on 145 million acres.
- Enter into nearly 313,000 (EQIP) contracts.
- Create or restore wetlands on 2.7 million acres.
- Apply comprehensive nutrient management plans on almost 40 million acres.
- Develop new or updated soil maps on 260 million acres.
- Deploy a new Web Soil Survey Program with more than 3.5 million website visits by the public.

These accomplishments are a testament to the continued trust and relationship that we maintain at the local level with farmers, ranchers, Conservation Districts, and other partners. As we initiate implementation of the 2008 Farm Bill, with its increased investment in conservation programs, NRCS looks forward building on these accomplishments.

Growing Conservation and Some Growing Pains – the NRCS Financial Audit

While the results of conservation programs and investments have reshaped the landscape, it is clear that just getting conservation on the ground is not the full measure of program success. With the change in the scope of conservation programs and expenditures, it has come a realization that we need to better assess and maintain excellence in accounting procedures and execution, and to ensure that our recordkeeping systems are robust.

In FY 2008, NRCS contracted with an external audit firm to conduct our first standalone financial audit, under the supervision of the USDA Office of Inspector General and the USDA Office of the Chief Financial Officer. At the end of the FY 2008 audit, the auditors issued a disclaimer of opinion. The auditors found problems with the accuracy and completeness of the FY 2008 financial information. In some measure, this was due to inadequate recordkeeping in NRCS offices. During the timeframe of the audit period, NRCS was unable to provide the auditors adequate support to verify our financial information as presented for FY 2008. In other words, we could not prove the validity of our numbers.

The auditors found five material weaknesses: accounting and controls for (1) undelivered orders, (2) unfilled customer orders, (3) accrued expenses, (4) property, plant and equipment, and (5) controls over financial reporting. They also identified deficiencies in our internal controls over purchase and fleet card transactions, and the general controls environment for our information systems.

NRCS understands the seriousness of these findings and is moving aggressively to correct them. When informed of the auditors' preliminary findings, NRCS began developing a corrective action plan and initiated a massive undertaking--a review of over 160,000 open obligations. To our knowledge, a review of this size and scope is unprecedented in the federal government. The agency developed and delivered training to over 330 NRCS personnel in mid-November, 2008 and continues to aggressively review open obligations. So far NRCS has deobligated over $1.3 billion since the review started in FY 2007. To help prevent this from reoccurring, NRCS now mandates that all line officers formally certify on a quarterly basis the accuracy, reliability, and completeness of information in 21 separate areas of financial management.

During this file-by-file, transaction-by-transaction evaluation, we learned a great deal about our existing contracts, easements, and other open obligations. As a result of the audit and our aggressive approach, we have outlined a comprehensive corrective action plan necessary to establish a firm foundation for going forward. NRCS is analyzing and rewriting policy and procedures for program, administrative, and financial aspects of our business to ensure that all responsible parties understand what is required. In addition, we have begun an initiative to redesign and streamline our business processes. I am confident this initiative will lead to the development of new strategies for delivering conservation assistance that are more efficient and effective.

The external auditor is currently performing a special review of corrective actions taken to date for the FY 2008 audit. The results of this review will be available in April. In addition, the audit firm has started work on the FY 2009 financial audit. Our goal is to have a clean audit in the near future.

Clarifying the Term, "Deobligation of Funds"

Prior to the stand-alone audit, a limited scope review in FY 2007 showed a high number of fund deobligations within our agricultural conservation programs. *Deobligation* of funding occurs when funding that was previously obligated—either through a contract or agreement—is released because of cancellation, termination, modification or spending adjustments.

A key point to remember is that whenever funds are deobligated, they are not lost to the taxpayer nor are the funds necessarily lost to a prospective farmer or rancher. Funds deobligated in our discretionary programs—Conservation Technical Assistance, Emergency Watershed Protection, Watershed Rehabilitation, for example—are generally shifted to other priority projects within the respective program. Funds deobligated in mandatory Farm Bill programs, if not used for contract modifications or cost overruns, are eventually returned to the Treasury.

There are a number of reasons why funds may be deobligated out of contracts. These reasons vary across the diverse suite of programs delivered by NRCS. Some deobligations historically have occurred because of how NRCS delivered its programs. Here are some examples:

(1) A WHIP contract included a plan for a field border, including the number of acres and the costs associated with creating the border. Both the number of acres and the costs were estimates at the time of obligation. Two years later, when the producer went to install the field border, the costs both came in less than estimated. The excess funding in the contract resulted in a deobligation of the difference between the estimate and the actual cost.

(2) For a WRP contract, restoration costs were estimated based on a preliminary restoration plan. When the wetland restoration was actually completed some time later, it was found that the restoration costs were overestimated, leading to deobligation of some funds.

Deobligations also routinely occur because of noncompliance caused by the sale or transfer of property, changes in agricultural operations, death or serious illness of participants, natural disasters, bankruptcies, and personal hardships. These factors cannot be anticipated at the time a contract is signed. Here are a few examples:

- EQIP contracts can be up to 10 years in length. A producer signed up in year one with a commitment to install a grassed waterway in year five of the contract. Funding was obligated for all of the practices in the contract at the time the contract was signed at the beginning of year one. In year three, the producer passed away and the family decided to sell the farm. The funds for the grassed waterway had to be deobligated.
- In 2004, a producer signed a contract that included an animal waste structure to be built in 2006. After Hurricane Katrina, the cost of construction materials skyrocketed. The producer was unable to afford his or her share of the cost to build the structure in 2006, and the funds were deobligated.

Again, deobligations due to these types of producer noncompliance cannot be anticipated at the time a contract is signed. We have a keen interest in answering the question—what is an acceptable rate of deobligation for the types of programs NRCS administers? A 2005 Economic Research Service analysis estimated that the average annual exit rate for farms is nine to ten percent per year. Our latest estimated exit rate for EQIP contracts is thirteen percent annually. The constantly shifting mosaic of conditions in the agricultural economy and industry as a whole and at the individual farm scale indicates that some level of deobligation is expected. That is not to say, however, that NRCS is not committed to reducing deobligations. We have embarked on a number of efforts to do just that, to reduce to the greatest extent possible the number and amount of deobligations due to NRCS business practices and program policies.

A key point to remember is that Farm Bill conservation program contracts are distinctive agreements., These contracts are a product of an individual farmer or rancher voluntarily offering his or her own financial resources toward a benefit not just for themselves but for the public writ large. NRCS manages hundreds of thousands of conservation program contracts. It is inevitable that, with some frequency, a producer's personal or financial situation will change over the lifetime of a contract. Our objective is to ensure that farmers can be good conservation stewards while maintaining productivity and profitability. Cancellation of conservation projects are a reality and, given the emerging economic climate, may increase in the near term.

Moving Forward

Looking ahead, we believe we are better positioned to handle the issues raised by the audit and fund deobligation statistics. Starting two years ago, NRCS began developing a number of new business tools and practices that will improve our financial management controls. This fall, we will introduce a business tool that will integrate easement contracts into our financial management system. Currently, we are reviewing every policy document produced by the agency to find ways to improve program delivery, tighten financial controls, and reduce fund deobligations. In 2008, the agency implemented a new WRP business model that will result in improved payment controls and fewer deobligations. Two other program policy changes—payment schedules and a payment inflation index—should also help reduce future deobligations. Finally, as I mentioned earlier, we have launched an initiative to establish a new vision for delivering our programs and carrying out the agency's core activities—conservation planning and the application of conservation practices—through a new business model and modernized workforce.

The audit has been a positive experience for NRCS in that it pointed out ways that the agency can achieve a higher standard in implementation of its programs. The issues that the audit raised are solvable and we have taken aggressive action to immediately address the deficiencies and weaknesses in our financial system. However, we recognize that these issues will not be solved overnight. Our corrective action plan details actions that will be implemented over the next year and beyond. NRCS leadership is evaluating options to address accounting expertise across the Agency and issuing strengthened policies and procedures governing business and financial management processes. In February 2009, the USDA Office of Inspector General concurred, without exception, to our planned actions.

We believe we are on the right track to be better equipped for success in financial management for the future. NRCS has evolved greatly over the last two years in our understanding of proper accounting for our financial resources. We have embraced the financial audit as a way to improve achieving our mission and stewardship of taxpayer assets. I want to reinforce that the audit did not show any instances of funds being misused or improper payments. We recognize that there are three critical aspects of the situation: human capital, processes, and systems. Our planned remedies to the problems revealed by the audit will address each of these critical areas.

CONCLUSION

Conservation programs have provided notable achievements in both conserving and protecting our natural resources. However, several existing and emerging environmental challenges will require needed attention. Efficient and effective delivery of USDA conservation programs could not occur without a strong working relationship between FSA and NRCS. The Agencies will continue to work to improve the delivery of program services and to ensure the environmental benefits are achieved in a sound fiduciary manner. We thank the Chairman and members of the Subcommittee and would be happy to respond to any questions that members might have.

CHAPTER SOURCES

The following chapters have been previously published:

Chapter 1 is an edited, excerpted and augmented edition of a Congressional Research Service publication, CRS Report for Congress Order Code R40763 Updated March 25, 2010.

Chapter 2 is an edited, excerpted and augmented edition of a Congressional Research Service publication, CRS Report for Congress Order Code R40692 Updated Decemner 28, 2009.

Chapter 3 - This remark is delivered as testimony before the Subcommittee on Conservatin, Credit, Energy,and Research, House committee on Agriculture dated Wednesday, March 25, 2009.

Chapter 4 - This remark is delivered as statement of Robert Stephenson Acting Deputy Administration for Field Operations Farm Service Agency, U.S. Department of Agriculture and Dave White Chief dated March 25, 2009.

INDEX

D

E

F

G

H

I

L

M

N

O

P

Q

R

S

T

U

V

W